Future Tourism in a Robonomic World

THE FUTURE OF TOURISM

Series Editors: **Ian Yeoman**, *NHL Stenden University of Applied Sciences, the Netherlands* and **Una McMahon-Beattie**, *Ulster University, Northern Ireland, UK*

Some would say that the only certainties are birth and death; everything else that happens in between is uncertain. Uncertainty stems from risk, a lack of understanding or a lack of familiarity. Whether it is political instability, autonomous transport, hypersonic travel or peak oil, the future of tourism is full of uncertainty but it can be explained or imagined through trend analysis, economic forecasting or scenario planning.

This new book series, The Future of Tourism, sets out to address the challenges and unexplained futures of tourism, events and hospitality. By addressing the big questions of change, examining new theories and frameworks or critical issues pertaining to research or industry, the series will stretch your understanding and generate dialogue about the future. By adopting a multidisciplinary perspective, be it through science fiction or computer-generated equilibrium modelling of tourism economies, the series will explain and structure the future – to help researchers, managers and students understand how futures could occur. The series welcomes proposals on emerging trends and critical issues across the tourism industry and research. All proposals must emphasise the future and be embedded in research.

All books in this series are externally peer-reviewed.

Full details of all the books in this series and of all our other publications can be found on http://www.channelviewpublications.com, or by writing to Channel View Publications, St Nicholas House, 31-34 High Street, Bristol, BS1 2AW, UK.

THE FUTURE OF TOURISM: 9

Future Tourism in a Robonomic World

Edited by
Stanislav Ivanov and Craig Webster

CHANNEL VIEW PUBLICATIONS
Bristol • Jackson

DOI https://doi.org/10.21832/IVANOV9837
Library of Congress Cataloging in Publication Data
A catalog record for this book is available from the Library of Congress.
Names: Ivanov, Stanislav, editor. | Webster, Craig, editor.
Title: Future Tourism in a Robonomic World/Edited by Stanislav Ivanov and Craig Webster.
Description: Bristol, UK; Jackson, TN: Channel View Publications, 2025. | Series: The Future of Tourism: 9 | Includes bibliographical references and index. | Summary: "This book envisions the future of tourism in an economy that is largely automated. An economic system that relies on robots, AI and automation technologies for the production of goods and delivery of services instead of human labour is known as 'robonomics'. The volume explores the challenges of transitioning to a robonomic tourism ecosystem"—Provided by publisher.
Identifiers: LCCN 2024013455 (print) | LCCN 2024013456 (ebook) | ISBN 9781845419820 (paperback) | ISBN 9781845419837 (hardback) | ISBN 9781845419851 (epub) | ISBN 9781845419844 (pdf)
Subjects: LCSH: Tourism—Automation. | Tourism—Technological innovations. | Tourism—Forecasting. | Hospitality industry—Automation. | Hospitality industry—Technological innovations. | Hospitality industry—Forecasting.
Classification: LCC G155.A1 F89 2025 (print) | LCC G155.A1 (ebook) | DDC 338.4/7910285—dc23/eng/20240514
LC record available at https://lccn.loc.gov/2024013455
LC ebook record available at https://lccn.loc.gov/2024013456

British Library Cataloguing in Publication Data
A catalogue entry for this book is available from the British Library.

ISBN-13: 978-1-84541-983-7 (hbk)
ISBN-13: 978-1-84541-982-0 (pbk)

Channel View Publications
UK: St Nicholas House, 31-34 High Street, Bristol, BS1 2AW, UK.
USA: Ingram, Jackson, TN, USA.

Website: https://www.channelviewpublications.com
X/Twitter: Channel_View
Facebook: https://www.facebook.com/channelviewpublications
Blog: https://www.channelviewpublications.wordpress.com

Copyright © 2025 Stanislav Ivanov, Craig Webster and the authors of individual chapters.

All rights reserved. No part of this work may be reproduced in any form or by any means without permission in writing from the publisher.

The policy of Multilingual Matters/Channel View Publications is to use papers that are natural, renewable and recyclable products, made from wood grown in sustainable forests. In the manufacturing process of our books, and to further support our policy, preference is given to printers that have FSC and PEFC Chain of Custody certification. The FSC and/or PEFC logos will appear on those books where full certification has been granted to the printer concerned.

Typeset by Deanta Global Publishing Services, Chennai, India.

Contents

Contributors		vii
Preface		xi

Part 1: The Future History of Tourism

0.1968	Diary of Samantha Smith *Craig Webster*	3
0.2068	Diary of Siri Wang *Craig Webster*	13
0.5	The Transformation from 1968 to 2068 *Craig Webster*	25

Part 2: Foundations of Robonomics

1	Principles and Drivers of Robonomics *Stanislav Ivanov*	43
2	Consequences of Robonomics *Stanislav Ivanov*	65
3	Solutions to the Challenges of Robonomics *Stanislav Ivanov*	85

Part 3: Robonomics and Future Tourism

4	Implications of Robonomics on Future Tourism: An Overview *Craig Webster and Stanislav Ivanov*	103
5	Future Trajectories for Automated Tourism and Hospitality Services *Ellis Urquhart*	114

6	The Automated Tourism and Hospitality Company of the Future *Stanislav Ivanov and May Kristin Vespestad*	133
7	Creating Experiences through Automation Technologies *Katerina Berezina, Lisa Cain, Katerina Volchek and Cihan Cobanoglu*	148
8	Development of Robot-Friendly Hospitality Facilities *Katerina Berezina, Olena Ciftci and Fernando Arroyo Lopez*	170
9	Sex, Health and Wellness: Considering the Future Potential for Robots and Human Relationships within Tourism Resorts *Daniel Wright*	190
10	The Sustainability of Tourism in Robonomics *Craig Webster, Fernando J. Garrigos-Simon and Yeamduan Narangajavana-Kaosiri*	209
11	Future Tourism in a Robonomic World: An AI-Generated Chapter *Stanislav Ivanov and Craig Webster*	223
	Conclusion: The History of the Future: Robonomics and Tourism *Craig Webster and Stanislav Ivanov*	256
	Index	258

Contributors

Fernando Arroyo Lopez received his PhD in Nutrition and Hospitality Management in the Department of Nutrition and Hospitality Management at the University of Mississippi. His professional experiences motivated his research to focus on topics regarding perceived value and virtual reality in the hospitality industry.

Katerina Berezina, CHTP, CRME, CHIA, is an Associate Professor and a Hospitality Management Programme director in the Department of Nutrition and Hospitality Management at the University of Mississippi. Dr Berezina's research interests are in the areas of information technology in hospitality and tourism, electronic distribution and revenue management. She serves as the Managing Editor of the *Journal of Hospitality and Tourism Technology*. Also, Dr Berezina assumes the roles of the Vice-Chair of the CHTP Advisory Council with Hospitality Financial & Technology Professionals (HFTP), Director of membership services with the International Federation for IT and Travel & Tourism (IFITT) and is a member of the Industry–Faculty Partnership Council with Hospitality Sales and Marketing Association International (HSMAI).

Lisa Cain is an Associate Professor in the Chaplin School of Hospitality and Tourism Management at Florida International University. Dr Cain currently serves as President for the SECSA-ICHRIE and serves as an Associate Editor for both the *International Journal of Consumer Studies* and *International Hospitality Review*. Her research interests fall within the broad topics of organisational behaviour and marketing, with an emphasis on understanding internal and external customer experiences. Specifically, she has published in the areas of work–life balance, substance abuse among hospitality workers, gender issues, technology and loyalty in the hospitality industry. She continues to develop research in these topics.

Olena Ciftci, PhD, CHIA, is a Clinical Assistant Professor of Hospitality and Travel Technology at the Jonathan M. Tisch Center of Hospitality, within the NYU School of Professional Studies at New York University. Her research interest is in information technologies in the hospitality industry, revenue management, consumer behaviour and research methods. Olena is the author of articles in academic journals, publications in professional journals and two book chapters. She has presented her research at international academic conferences. Olena serves as an Editorial Assistant at the *Journal of Hospitality and Tourism Technology* and a reviewer for five highly ranked academic journals in hospitality and tourism.

Cihan Cobanoglu is the Dean and the McKibbon endowed Chair Professor of the School of Hospitality and Tourism Management (SHRM) in the Muma College of Business at the University of South Florida (USF), and also serves as Director of the M3 Center for Hospitality Technology and Innovation and Coordinator of international programmes for the School of Hospitality and Tourism Management. He is a renowned hospitality and tourism technology expert. Dr Cobanoglu is a Fulbright specialist commissioned by the Fulbright Commission (2018–2021). He is a certified hospitality technology professional (CHTP) commissioned by Hospitality Financial & Technology Professionals (HFTP) and the Educational Institute of the American Hotel & Lodging Association (AHLA). He is the Editor of the *Journal of Hospitality and Tourism Technology* (JHTT) (indexed in SSCI) and a co-author of more than 10 books and over 20 conference proceedings.

Fernando J. Garrigos-Simon has a PhD in management and is a Full Professor in the Department of Business Organisation, Universitat Politecnica de Valencia, Spain. He has been a visiting professor at universities in Australia, Bolivia, France, Finland, Germany, Ireland, Singapore, Taiwan, Thailand, the UK and the US. His research is related to tourism management, strategic management, IT management and crowdsourcing. He has developed international projects and published in international books and journals such as *Annals of Tourism Research, International Journal of Contemporary Hospitality Management, International Journal of Technology Management, Journal of Knowledge Management, Small Business Economics, Management Decision, Tourism Management* and *Tourism Economics*. He is a member of the editorial board of various international journals, founder of the conference INNODOCT and editor-in-chief of the *European Journal of Studies in Management and Business* and *Studies in Education Management*.

Stanislav Ivanov is Professor at Varna University of Management, Bulgaria and Director of Zangador Research Institute. Professor Ivanov is the founder and (Co-)Editor-in-chief of two academic journals: *European Journal of Tourism Research* (http://ejtr.vumk.eu) and *ROBONOMICS:*

The Journal of the Automated Economy (https://journal.robonomics. science). His research interests include robonomics, automation in tourism/hospitality, the economics of technology, revenue management and political issues in tourism. For more information about Professor Ivanov, please visit his personal website: http://www.stanislavivanov.com.

Yeamduan Narangajavana-Kaosiri is a Lecturer at Universitat de Valencia, Spain. She has a PhD in Marketing, Universitat Jaume I (with distinction), Spain; an MSc in Tourism Management and Marketing, Bournemouth University, UK; and a BA in Business Administration, Chiang Mai University, Thailand. She has been coordinator of the tourism department for the faculty of business in Walailak University, lecturer at Dusit Thani College (Thailand), Universitat Jaume I (Spain) and visiting lecturer at Bradford University, UK, UPV, Spain and Oulu University of Applied Sciences, Finland. Her research interests are tourism and marketing, prices analysis, social media and user-generated content. She received the award for the best research project of all University Jaume I in 2006. She has developed international projects and published papers in books for editorials such as Springer, Pearson and Routledge, and for journals such as *Annals of Tourism Research*, *International Journal of Contemporary Hospitality Management*, *Journal of Travel Research* and *Tourism Management*.

Ellis Urquhart is a Lecturer and Postgraduate Programme Leader in Tourism Management within the Business School at Edinburgh Napier University, UK. He specialises in visitor attraction management, co-creative experience design and technological mediation in the heritage sector and the wider attraction environment. Ellis teaches tourism management at both undergraduate and postgraduate levels at Edinburgh Napier University in addition to overseas programmes delivered in Switzerland, Singapore, Macau and Hong Kong. He currently reviews for a range of tourism publications, has appeared at a number of internationally recognised conferences and contributes to various academic publications.

May Kristin Vespestad, PhD, is a Professor of Tourism Marketing at the School of Business and Economics, UIT, The Arctic University of Norway. Her main teaching and research areas are within tourism marketing, tourism experiences, adventure tourism, consumer behaviour and experience co-creation. She studies, among others, marketing aspects and consumer behaviour in areas such as nature-based tourism and adventure tourism. She has published in several international journals and serves on editorial boards.

Katerina Volchek is a Professor and a Manager of the DigiHealth & Smart Tourism lab at Deggendorf Institute of Technology. She started her career as a product manager and marketer for a tour operator. Katerina is now an expert in customer experience, information and communication technologies and marketing strategy for tourism, including the design of personalised services and the optimisation of ROI through marketing attribution. Currently, her research interest also lies in the capabilities of neuromarketing and smart environments for the service industries. Katerina serves as Director for marketing at the International Federation for Information Technologies and Travel & Tourism (IFITT).

Craig Webster is a Professor of Hospitality Innovation and Leadership in the Department of Applied Business Studies at Ball State University, IN. He studied Government and German literature at St Lawrence University, received an MA and PhD in Political Science from Binghamton University, and an MBA at Intercollege, Cyprus. He has taught at Binghamton University, Ithaca College, the College of Tourism and Hotel Management in Nicosia and the University of Nicosia. His research interests include the political economy of tourism, robots and artificial intelligence in service industries and public opinion analysis. Dr Webster has published in many peer-reviewed journals and is co-editor of two other books. He currently teaches courses in hospitality management at Ball State University's Miller College of Business.

Daniel Wright has published widely around tourism futures in the academic literature. He uses futurology as a transdisciplinary field of study to forecast, anticipate and provoke the future of tourism in global visitor economies. Daniel is a member of the Institute for Dark Tourism Research (iDTR) and is an editorial board member for the *Journal of Tourism Futures*.

Preface

> Change is one thing, progress is another. 'Change' is scientific, 'progress' is ethical; change is indubitable, whereas progress is a matter of controversy.
> Bertrand Russell, *Philosophy and Politics* (1947: 14)

The world is experiencing a phase transition. Automation technologies are becoming ever more pervasive. Robots produce goods and deliver services. Autonomous vehicles transport passengers. Chatbots communicate with customers, make bookings and deal with complaints. Implantable devices collect biometric data. Generative artificial intelligence (AI) applications create on-demand content (text, video, sound, code). Autonomous agents trade on the financial markets. AI decides what we see on social media platforms. Humans hasten the adoption of automation technologies with a severe demographic crisis caused by low birthrates. In the future, all these technological advancements and demographic developments will accelerate to a point where humans will no longer be needed in the economy. Such an economic system that relies on robots, AI and automation technologies for the production of goods and the delivery of services instead of human labour is known as 'robonomics'.

This book investigates how tourism will look in a future, highly automated, technology-reliant robonomic society. It is divided into three parts. Part 1, composed of Chapters 0.1968, 0.2068 and 0.5, looks at the technological, demographic, economic, cultural, political, environmental and other changes between 1968 and 2068 through the eyes of the tourist. It serves as a contrast, showing the ways that robonomics will transform the tourism experience of consumers. Part 2 (Chapters 1–3) lays the theoretical foundations of robonomics as an economic system – its characteristics, drivers, benefits, challenges and potential solutions to challenges. Finally, Part 3 (Chapters 4–11) elaborates on how the robonomic economic system will impact on future tourism. In particular, it delves into the future trajectories for automated tourism and hospitality services, the automated tourism and hospitality company of the future, creating tourism experiences with automation, the robot friendliness of hospitality facilities and human–robot relationships in the tourism context. In the concluding chapter, we investigate the perspective of AI itself by interviewing ChatGPT on future tourism in a robonomic world.

Although it discusses technologies at length, this book is not about technologies but more about the relationship between technologies and humans. Our philosophy is that technologies are tools that allow humans to achieve certain goals. Humans are at the centre of the book – tourists, tourism and hospitality employees and their families, local residents at destinations, managers, policymakers, administrators, software developers, engineers, marketers and all others involved directly or indirectly in tourism and hospitality. The chapters discuss the benefits and challenges of a robonomic economic system and tourism through the lens of *human* stakeholders. A future robonomic world will allow people to live longer and healthier lives largely liberated from the stresses of work, but it will raise issues related to social instability and unrest. Societies will be reorganised to reflect changes in the economic system and sources of income and economic power. Jobs will be eliminated or transformed, while new jobs will be created. People, the constituent element of societies, will have to adjust to new technological, social and economic realities. Transition will be difficult for many but ultimately beneficial for most. Part 2 sheds light on this process by explaining the characteristics of the system, exploring the demands for new institutions and the likelihood of turbulence in the transition.

In the future, tourists will still travel outside their usual environment to various destinations. While the core concept of tourism will remain the same, the way tourism is produced and consumed will evolve. The future of tourism, as with almost all service industries, is automation. Regardless of whether or not we want the automation of services, and independent of our personal attitudes towards automation technologies, numerous factors (mainly demographic and economic) are elaborated on in the book that work in favour of greater levels of automation. The confluence of a declining human workforce and massive increases in technological capabilities leads to a future in which automation is the norm in virtually all industries.

In tourism and hospitality, robots will cook in restaurants, provide information and register guests in hotels, and transport tourists. Artificial autonomous agents will determine prices, buy from suppliers and authorise invoice payments. Some tourism and hospitality companies will be completely automated and may have no human employees, being run entirely by AI. While tourism is often considered a people's business where humans serve humans, the implementation of automation technologies in tourism and hospitality is not necessarily bad for employees or consumers. By automating dirty, dull, dangerous and repetitive tasks, especially those in back-of-house operations without customer involvement, automation technologies allow human employees in tourism and hospitality companies to spend more time with guests. Consumers are also expected to benefit from lower prices due to economies of scale and technological advancements. Moreover, technologies will facilitate the hyperpersonalisation of tourism experiences; hence, it may turn out that future tourism may be more guest-centric than current tourism despite the technological shield between the tourist and the tourist company.

The chapters of the book will immerse you in the experience of tourism in a robonomic world, a world that is moving from fantasy to reality at a rapid pace. They will raise many questions, provide answers to some of them, but mostly elucidate what is needed to prepare society for the massive changes coming. They will look at the likely trajectory that societies and economies are going to take considering the technological advances. However, we should not forget that the future is not predetermined; it is yet to be created.

Let's co-create a brilliant future together!

Part 1
The Future History of Tourism

0.1968 Diary of Samantha Smith

Craig Webster

(Photo created by the authors using ChatGPT)

June 7, 1968

Maybe it is not a good idea with how crazy things have been, I want to see Europe. I am glad that we bought the tickets to see Europe. I went to the travel agent before the weekend. I think I have enough money saved up so that Bob and I can travel to Europe. While we have not had luck making children yet, we have some money saved up so we can afford it. The past few months have been crazy with the massive fighting in Vietnam, the killing of Robert Kennedy two days ago, and the assassination of Martin Luther King just a few months ago, I want to get away. Fortunately, none of my friends from high school seem to have been killed in Vietnam but I know that many more have. It did not take so long at the travel agents to get the tickets and it was not as expensive as I thought it would be. I am a bit excited because I have never flown abroad before.

It will be an adventure and I look forward to it. We are going to fly to London! It will be a long flight, something none of our parents have done. It should be fun, especially because we have to make a stop in Gander for refueling, so I guess I am also lucky to get to Canada on the way, even if it is only for a short stop.

July 10, 1968

Tomorrow is the big day! We are going to Europe. I packed our bags and am ready to go. Bob and I are very excited about traveling to Europe. He is especially excited because the last time he went, he had to take the boat to Europe and that took way too long. He is happy to fly because it will be much shorter than when they sent him to France eight years ago and he told me about how sick he got in the boat on the way over there.

July 11, 1968

We made it! Roger drove us to Kennedy Airport. He is a good brother but a terrible driver! Since it was my first trip abroad, I was really excited about this whole process. I went to the counter and showed our tickets and passports. They took our bags and we walked into the terminal to go to the plane. Bob and I did not have to wait a long time. We were both excited and talked a great deal about what we were going to do in Europe.

July 12, 1968

We made it to London but it was a bit exhausting! We arrived in London and we are really happy about the trip. The plane was a bit louder than I thought it would be. The pilot spoke every now and then to remind us of where we were and how much longer it would be to make it to London. During the flight, we had some food that the stewardesses brought to us and I tried to sleep as much as I could. In Gander, I got out and absolutely loved the ice cream they have there. Bob talked about how much better it was to be in the airplane for just a few hours than being on a ship for more than a week. But we were both really happy to get out of the plane in London.

As soon as we arrived at the airport, we got out of the plane and went through customs. It was not so bad, we had to show our passports and everyone was really nice. But when we got out, we were really tired from traveling but had to find a place to stay for the night. We took the train to downtown London and found a bed and breakfast. It took some time but we finally found it. We went to our bed and breakfast, the room was small but I was grateful just to have a place to sleep and the bathroom was just down the hall a little bit. It was a small house and it was not too far from all the places we wanted to get to, so we were happy. I wanted to sleep so badly but Bob told me we should not because we would have all sorts of problems with jet lag.

July 13, 1968

It's our first full day in London and I am excited to see so many things. We saw Parliament, Big Ben, and the Tower of London. Growing up, I learned so much about how the people in London put up with the daily bombings from the Germans. Looking around, it is amazing to think that less than 30 years ago, this city was being bombed day after day.

I guess there are some similarities between London and New York, my home. They are big cities and have lots of subways. The subway system here is a bit confusing, and buying the right tickets to go to the right place takes some time and energy to figure out.

A surprising thing that is hard as a tourist abroad is figuring out the money. We brought plenty of money in traveler's checks but it is not always easy to use them. We did go to a bank yesterday to change money. The hard part is that we have to figure out how much money a person needs to use because we don't want to change more money than we need and we want to make sure we have enough money for the trip. So it was important for us to make it to the bank before the weekend because we wanted to figure out how much money we needed for the next few days.

Figuring out the money alone is a bit hard because the money in this country does not seem to make sense. There are 12 pence in a shilling and 20 shillings in a pound, which means that there are 240 pence in a pound. For those of us who come from a country with a hundred pennies to a dollar, the math that a person has to do to understand what he is paying is very confusing. But that is part of the fun of traveling, I guess.

Part of the fun too is that the people really do dress differently here. I never really imagined that I would see so many people walking with bowlers on the streets. But I suppose that this is just a general difference. But then, I suppose that the people here probably think that Bob and I dress strangely. It seems that Americans have a different sense of style and clothing than Londoners.

We are starting to feel a bit better from the jet lag today. I hope that by tomorrow we feel less dopey from the jet lag.

July 14, 1968

Today we are feeling better and we have fully recovered from our jet lag. It is Sunday so many things are closed. We were lucky enough to make it to the museum. The British Museum was really something amazing. It was nice to go and get a ticket and see the museum. The man at the ticket booth was so helpful, telling us which exhibits would be the best for us to see and marking on the papers where we should go to see the best exhibits in the museum. Because he gave us such good advice, we saw things that I will tell all my friends about for years to come. We saw the Rosetta Stone and some of the sculptures from the Parthenon. It was a good experience. We walked around and read all the information available on the walls and in the exhibits. We learned so much about history and really enjoyed the museum. Next time I am in London, I want to go to the Natural History Museum, as well.

When it came time for lunch, Bob brought me to a place to get some fish and chips. They served the fish and chips in newspapers and we had a good time eating them with lots of malt vinegar. I don't know why but I always loved the way that fried fish smells. It was delicious and we talked with the people in the restaurant about how we came to visit and wanted to see more of the city. They gave us some advice about other things to see in the city but Bob said that we should keep on traveling and get to see some more things in France.

I think Bob is right. There is more to see in France. Too bad we are in London for Bastille Day. I bet they are having some serious celebrations in France for the holiday. But we are on the other side of the Channel.

July 15, 1968

We got up in the morning and went to the train station to get to Dover and get over to Paris. It was not as easy as it sounds. We had to go to the train station and get the earliest train we could. We had been told that there was a morning train but we did not make it in time because we were delayed, since it took us longer to buy the tickets at the counter in the train station than we thought. So, we had to wait in the train station for the next train which was just an hour or so later. But we did not mind, we got a newspaper and sat and waited at the train station.

The newspapers keep reporting on all the positive things that are happening in Czechoslovakia. It is funny that we are on the continent where so much has happened and not far away from Czechoslovakia where so much is happening now. I sure hope that things continue to improve in Czechoslovakia. I know that some good things are happening in this world, since there are supposed to be commercial flights from the USA to the USSR as of today. That is something that would be hard for many of us to imagine but it is happening.

The train to Dover did not take very long. For an American, it reminds you that Britain is very small compared to the USA. The countryside was green and pleasant to look at. What was fun is that we sat in a train compartment with six others who were also going to Paris. They were all British and going to the continent for some adventure. While it did not take long to get to Dover, it took some time to get out and board the ferry to France. Fortunately, the waters were pretty calm and it was only about two hours to cross the English Channel to Calais in France. It makes a person respect the people who are hellbent on swimming across the Channel since it certainly is longer than I would ever be able to swim.

Once we arrived in Calais, we had to go through customs. It is good that I am not too young and do not look too subversive since they seem not to like people who look like trouble. I guess since all the trouble in Paris in the spring, the police were concerned about what types of people are coming into the country. They did not seem to be too concerned about me and Bob so they let us into the country without much trouble. Of course, we had to show them our passports and were asked some questions about where we were going and what we would be doing in France. But it seemed like we were not the sort of people they were really worried about.

In Calais, we boarded the train for Paris. Before we boarded, though, we were lucky enough to pick up some sandwiches to bring with us for the trip and an Orangina, which is an orange soda. It may not be my favorite soda in the world, but it is a bit different from American sodas with some fruit bits in it and a very stylish bottle. It is sort of interesting how even the bottles for sodas are different in France. They sure have a sense of style, and I hope some of that stylishness can rub off on me. The cheese in the sandwiches was something I was not really familiar with. Bob said it was brie, and it was very soft and creamy.

It did not take too many hours to eventually arrive in Paris. But the next time, I think I would recommend taking the night ferry and sleeping in the train because the trip took so many hours and ate up so much time in the day. The good news is that we got the chance to speak to some people in the train and got to see the countryside. It was hard to speak to many of the French folks in the train. We were lucky because the personnel working in the train could speak some English but anyway all we really needed to do was to show them our tickets.

When we arrived in Paris, it was late but we were able to find a hotel near the station. It was a bit of an annoyance to walk from hotel to hotel with our bags to see if we could find something vacant. There was no room for us in the first hotel but we got lucky and there was a room for us in the second one. The hotel was small and comfortable. The smell of detergents that they used to clean the floor was memorable to me and is different from the cleaning solution brands we have back in the USA. We were also lucky because in this hotel, there was a bathroom in our room. The owner seemed nice and spoke some English even if most of his staff could not. We paid in travelers checks and went to bed. But after a long day of travel, I just wanted to sleep.

July 16, 1968

It was our first full day in the City of Light. Breakfast was some bread, croissants, and coffee. But the croissants were buttery, flaky, and delicious. The coffee was served with milk in a big bowl, something I had never seen before except in a French movie once.

Paris is really beautiful and it is like you can hear accordion music in the streets, I think you can actually hear it in a lot of places. Although you will also hear traffic and horns honking in many places, as well. We walked in the Latin Quarter, saw the Eiffel Tower, and went to Montmartre. We could still see that there was a lot of cleaning up to be done in the city since the riots and demonstrations in the spring made things a bit unpleasant. Workers were repairing the streets where the cobblestones had been ripped out of the street to be thrown at police. I am glad that I came after this and was not here while all hell was breaking loose.

I have to say that Paris really is something else. There are cafés on the streets and fantastic bakeries. The food culture is something. What I liked about this country is that I finally understand the money. Once we went to a bank and got some local money, we were able to figure it out really quickly. One hundred centimes make a franc, that was not hard to learn. The bad news is that we have a pocketful of coins from London that we don't know how to use here. I guess that it will be a good souvenir. But the bad news is that Bob has a trick for dealing with the waiters at the cafés – he hands them a big bill of francs to pay for the food and drinks so they think he speaks French. None of the waiters seem to speak English, so that makes ordering food hard because we are not always sure of what we are ordering. I have accumulated so much change that I regularly give the coins to the musicians on the streets, at least that is one way to deal with my problem of having too many coins.

I am a bit too tired to write much more today.

July 17, 1968

This is a big day for us, we went to museums. Of course, I could not miss the Louvre. I went there with Bob to see all the amazing art. We purchased a ticket but it was a bit hard to navigate the museum without a guide and without a really good idea of what we could see. None of the staff there seemed to be able to help us since there was a language barrier. Anyway, we saw the Louvre and spent many hours there really enjoying the art. We were not really sure where to go to get something to eat when we were done, since we did not see a restaurant or café immediately outside. But eventually, we found a place and had a great dinner. Bob brought me to an Alsatian restaurant, so the food looked really German to me with sausages and sauerkraut. But we had wine with the food, even if it looked German to me because the French wine is so good.

July 18, 1968

This is our last day in Paris. We went to Versailles and it is really impressive but it was not easy to get there. We took some buses and Bob could help with that with his rudimentary French. The whole palace is really something to see and it was much quieter there because we were away from all the traffic of Paris. But it seems like we really have to be moving, since we don't have many days left and there is still so much to see.

July 19, 1968

We took another day to see some of the sights in Paris. We visited another museum and walked the streets of the Latin Quarter. It was our last day in Paris and we were ready to leave for the coast. Bob tells me that the Riviera is really nice and I want to see it, since I just know it from the glamour of Cannes and Nice in the movies.

We checked out of the hotel and took the night train to Nice. It is good that we bought the tickets yesterday at the Gare d'Austerlitz and it is my first time sleeping in a train. It was sort of interesting getting into a train with Bob and some strangers came into the sleeping car with us. It was an experience sleeping with four other people who were strangers and we could not really communicate well with them, even if we all saw each other in our underwear. It was important to make sure we got in the right train and the right car. I look forward to waking up on the French Riviera.

July 20, 1968

I am definitely not regretting being on the French Riviera. The fresh salt air is such a pleasant change from the city air and the occasional smell of pee outside of the pissoirs on the street. We woke up in the morning in the train station with the conductor telling us to wake up since we are in Nice. It is a very different environment than Paris. I waited at the train station while Bob went to find a hotel. We were lucky because he found a hotel room not too far from the station. Again, we were lucky and found a hotel, even if there was no bathroom in the room and we had to share a bathroom with other guests down the hall. They have a strange thing in the bathroom in France that looks like another sink but it is used to clean up after using the toilet. They call it a bidet and it seems so strange to me. I am lucky that Bob knew what it is.

Even if we do not have a view of the Mediterranean, we are close enough for a short walk to the beach. We had lunch and checked in. There were some problems ordering food, since the menus were only in French but we have started to figure out some of the language although it takes some time to translate some of the language on the menus with the help of our dictionary.

It is warm and sunny. We will enjoy a few nice days on the beach.

July 21–24, 1968

I am spending many days at the beach and I am enjoying it. There are many young women who do not wear tops at the beach. I spoke to some people at the hotel and it seems that this is happening more at the beaches since all the protests in Paris in the spring. Bob and I rented a car to go to different beaches and we really enjoyed it. It was not too hard to rent a car but it was a bit hard for us to figure out where we were going. What really made it easier was that the Mediterranean is on one side (the South) so it helped us a lot with navigation. Many of the beaches are quite rocky but it was nice to be in the sun and see where the Europeans spend their vacations. It is not always fun to walk on a rocky beach but it is nice to get some sun and swim a bit.

We did go to a beach in Cap d'Agde and there were so many Dutch and German tourists there. While the French women seem to be willing to take their tops off at the beaches in Nice and Cannes, it seems that the Germans expect to have nothing on at the beach. No one seems to really care and I wonder how legal it is but no one seems really shocked by it.

July 25, 1968

It is time for us to leave the beaches and get ourselves back to Rome for the flight back home. It is a bit unfortunate since we have just been able to figure out the menus at the cafés and restaurants at this point. But now, it is time to move to another country and another currency and language. We still have more French money than we thought we would need. So we will see if we can change some of it in Italy or use it on the train if there is a restaurant car.

The border really was no trouble. We just had to show our passports and they did not ask many questions.

We think we have enough time to see some things in Italy before we return home. We took today as a travel day to get to Milan. I wanted to see the city there because I know the cathedral there is impressive. We purchased some tickets a few days ago for the trains to get us to Milan. Most of the day we traveled. It was good to see the countryside. It is a pretty but mostly dry countryside. We finally did make it to Milan where we found a hotel right next to the train station. We made it to a bank in good time to get lira and using another currency is a bit confusing, especially since we are spending thousands all the time.

July 26, 1968

It was not as exciting a city as I had thought it would be. We saw a lot of the city and decided to take the night train the next day to Rome. Bob and I struggled a bit with the people at the train station to get tickets for the overnight train to Rome. There was a language barrier but we eventually figured it out. So we jumped on the night train to get to Rome. Sleeping on the train is not the most comfortable experience but it sure is an adventure. It is not every day a person strips off their clothes down to their underwear and sleeps in close quarters with five other people.

Bob and I had some food with us and were happy. I was not able to find Orangina in Milan so I had to buy some other drinks that I could find in Milan with us. Fortunately, our bunkmates offered us some wine that they had. It seems to be a tradition to share food and drink in the train with those in the compartment.

July 27, 1968

Finally, we made it to Rome! And it is really hot here. Fortunately, Bob and I planned ahead and changed enough travelers checks to get enough money to get us through the weekend. Again, Bob left me at the station to figure out where the hotel was. He had booked it a few days before when we were in France. Although he had booked it, they were still a bit surprised to see us coming and there must have been some confusion about the room we booked. Fortunately, we had a room with a bathroom in it, so we did not have to share with others. The hallways of the hotel smelled just like the hotel in France, so they must use the same detergent for mopping the floors.

What I noticed is that in Italy people dress differently than in London and France, they have a whole different sense of style. We got a taxi to carry us and our bags to the hotel, since it was a bit further from the train station than I had hoped it would be. If only our bags had wheels on them.

July 27–August 1, 1968

Because the hotel was not too expensive and because there is so much to see and do here, we decided to stay until we fly on August 2nd. To see more of the city, Bob and I found a tour guide that the hotel recommended. They brought us on a bus and gave us a tour of the city, explaining the many different historical things that we could see and do here in Rome. We thought that would be a good way to learn about the city. The tour bus drove us around and the guide explained in different languages what we were looking at, not that we could understand her because the sound system was so bad. It was good that we also had our tourbook with us to give us maps and information to learn about what there is to see and do in the city.

The city of Rome really is like a museum. Everywhere you look you see history and lots and lots of Roman ruins. This is something very impressive as a person growing up in the New World. What is surprising is that it is just laying out there in the open and unprotected. So everywhere there is a park or empty space, you see Roman ruins.

Anyway, the food has been fantastic, even if we are not always sure what we are eating. Bob saw that the lunch in a restaurant was 'lingue'. Because it was the special, he suggested we all order it. It was a big piece of tongue on the plate. I had never seen something like that and it really turned me off, a big steamed tongue on a plate. I tried to eat some but it was not easy. The good news is that the fruity Italian wine washed down the food quite well.

We have been enjoying our last days by not traveling too much and just enjoying the city, seeing the sights by following the map in our guidebook. We do occasionally find people who work in hotels or at some of the sights who speak English. We don't want to try to speak in German in many places, since my high school German may not help much and many people here may have bad memories from the war.

August 2, 1968

We slept well last night and needed to sleep well, since it will be a long trip ahead of us. We were able to speak to the hotel, and they ordered a taxi to take us to the airport. We had to make sure that we had enough lira to pay for the taxi.

We got to the airport and were glad that we had not forgotten to take our tickets with us because sometimes Bob forgets to bring things. It was a pretty simple procedure. The taxi dropped us off at the airport and we carried our bags to the desk where we checked in. After checking in, we had to go to the gate and wait.

It was a great trip and I look forward to getting home with a handful of coins in my pockets that will be a good souvenir of this trip.

0.2068 Diary of Siri Wang

Craig Webster

(Photo created by the authors using ChatGPT)

June 1, 2068

This was an interesting day. Cleaning out grandma's things, I found she had stored her mother's diary in a dusty corner in the house. I guess my grandmother was a real pack rat, keeping things forever. Writing diaries may seem pointless but it seems that it has become some sort of a family tradition. I found it really interesting to learn about my great-grandmother's life in 1968. I guess so many things have changed since then. I notice that she had a really interesting life and that so many things seem to have been different in those days. It is funny that she was about the same age and married, just like me. I thought that it would be fun to retrace her footsteps on a European vacation, since I have some time and money to do it with it.

I think that my great-grandmother would have been surprised to see that her written words inspired me so much. She would have been shocked, probably, to see that right after reading her diary, I asked Yukiko, my new digital assistant, to look into recreating the European trip for me and George. Yukiko did a stellar job, looking into flights, accommodation, and things to do in each of the destinations that Great-Grandma Samantha went to. It is funny because Yukiko even knew that I would want to leave on the same day as my Great-Grandma Samantha. Yukiko knows me well. But then it seems Yukiko made the recommendations also based on biometric scans and my working hours and therefore knows when to best schedule a good vacation for me. Yukiko has been reminding me to take a break for a few weeks but I kept putting it off, since I am busy with work and have been busy cleaning out grandma's belongings.

I guess that I am ready to do a trip like that but I will have to think about it for a few more days, if I can convince George to do this with me. To help me out, I asked Yukiko to communicate with George's digital assistant to organize his schedule so we can both take the trip together. I asked Yukiko to make a plan and to surprise me once or twice.

I think it would be a good thing to do. It would be really fun before we begin the procedures to get me pregnant in September. He and I have designed the baby we want, have received government permission for the pregnancy, and are planning to do it the old-fashioned way with the embryo embedded in September.

June 4, 2068

I guess I am getting too anxious and really want to go to retrace Great-Grandma Samantha's steps, so I am glad that Yukiko reminded me that price changes were happening so George and I would have to make the bookings now. I asked Yukiko to book the flights, trains, and hotels in the destinations that Great-Grandma Samantha went to. I told Yukiko not to buy museum tickets because I wanted to see what the weather would be like to visit museums on rainy days. Since the weather forecast is not so specific that far out in advance, I wanted to check when it gets closer to my travel time. Yukiko worked with George's digital assistant using what they know about our likes and dislikes for hotel bookings and travel and kept us within our budget.

This is a good time for me to travel. Although I have already traveled a lot this year and I have to pay the tax for traveling more than 20,000 km, I have worked for the past few months, so I can afford the tax. My project is ending soon and I won't be working for the next few months, probably. Since I am way over the universal basic income, I will not get more money from the government this year and I have enough saved up from when I was working, so I can go and spend some time away from home.

After dinner, I talked with George to verify that we would go ahead with the trip that Yukiko had set up for us. Even though I have already traveled a few times abroad this year, this trip will be a bit special, since it will be a retracing of the steps of my great-grandmother.

July 10, 2068

Tomorrow is the big day; we are going to Europe. Yukiko took care of the tickets and security, so all I have to do is get to the airport with George and fly. I also like that Yukiko had checked to make sure that our passport and immunization records were up to date so we would have no issues on the trip. I did make sure that my bags had everything I needed. Yukiko filled me in on the weather in Europe, so it was nice that I was reminded to bring something for an occasional rain. Although we had been to Europe a few times together before, this will be something different since we will retrace the trip of an ancestor and we have generally avoided Paris and London before because George tends to travel more to Germany for his work. This will be a bit different for us, seeing some different cities.

July 11, 2068

I woke up in the morning and had a good breakfast, excited about the trip. While I have been abroad many times before, this would be a special trip and probably the last time we travel with just the two of us for the next few years. Fortunately, Yukiko ordered the self-driving taxi a little earlier than I would have, since she knew that there was going to be a bit more traffic than usual on this day. No matter what, there always seems to be heavy traffic on Long Island. I liked that George and I could enjoy the last episode of a series we were watching when we were in the taxi. Although the automatic driver asked us a few questions on the way (about preferred air temperature and something about any additional information needed), we could mostly just enjoy the last episode in peace.

It was an easy ride to the airport, the taxi brought us, and the robots at the airport took our bags and delivered them to the check-in for the flights. As usual, Yukiko paid for the taxi and robots to carry the baggage. At the check-in, the robots processed the chips in our arms and put the luggage into the right system to be loaded onto the plane. I am glad that I have these embedded technologies because I do remember my mom telling me a story about how she had her airplane tickets and passport stolen from her.

George and I walked out to the terminal. Since we are a bit early, we could sit down and have a coffee. The automat gave us a good but somewhat overpriced coffee and we waited, talking about what we would do in London and Paris.

We jumped in the plane and in a few hours arrived in London Heathrow. The flight was pleasant, even though there was some significant turbulence on the way. I guess the climate change of the past few years makes these things a bit unexpected, even if we know that there is more turbulence in the atmosphere.

Fortunately, Yukiko had sent my food preferences to the airline, so that I did not need to make any special orders for food and the delivery robot brought what I like to eat. The robotic pilot gave us all updated information about the flight, telling us when to prepare for turbulence, and occasionally letting us know if our arrival time had been adjusted. Because he likes to fly, Yukiko reserved the front seat in the plane so we could have a good view. In the old days, the pilots sat way up there but we were lucky to have the view on this flight. It is good that we don't need human pilots anymore and that we get such good seats.

Arrival was easy, the robots gathered our baggage and we went through passport control quickly. We had to have our chips scanned and they verified who we are. There was a funny experience, though, since George's twin brother had been in London a few days before, so the facial recognition technology misidentified him. A person from security came down and the confusion was settled quite quickly, once we explained the issue. This has happened to George once or twice before but it always makes for a fun story.

As always, the porter robot took our baggage to the taxi station where Yukiko had ordered and paid for our taxi to the hotel. It was nice to have Yukiko find the taxi and hotel that were right for our needs and pay for them. I was glad to get to the hotel near downtown London. It was late in London and I knew that tomorrow I would have jet lag. So, it was good that security could scan us and perform the check-in and inform us about where our rooms would be. We took the elevator to the room with the hotel's porter robot that took our bags to the room. We were scanned again and the door opened for us. It's been a long day! I was glad to get to a comfortable bed that our digital assistants had requested for us. The scanners in the room dimmed the room. We were not very hungry, but the 3D-printer made a small snack for us.

July 12, 2068

It was good to wake up in London. Upon waking up, Yukiko communicated to me the messages from the guest relations manager at the hotel and told me what promotions and new offerings were available in the hotel and spa center. I was too excited to see the city to take advantage of many of the offerings at the hotel, so I told Yukiko a bit about my intentions for this travel, so she would not tell me more about spa promotions, unless there was a rainy day coming up or if she found I was a bit stressed or tired.

George and I went downstairs to see what the breakfast looked like. We saw the robots in the hallways silently cleaning the floors, and we noted that the detergents used for cleaning smelled different than the ones back home, something my great-grandmother noted so many years ago. Some things never change!

The hotel had the standard breakfast fare for a buffet. There was the typical European continental food, baked beans, and eggs, but also the usual Asian foods (steamed buns, noodles, and tofu pudding). I liked the tofu pudding and had that, since it reminds me of my grandmother on my dad's side of the family. As usual, we were scanned at the door to make sure that we got the food that was part of the package with the room.

Since we were awake early and it was rainy, we decided to go to the British Museum, based on Yukiko's suggestion for the day. I asked Yukiko to buy the tickets and order a taxi to get us to the British Museum. We left the hotel and our taxi was there immediately and brought us to the British Museum in no time. I asked Yukiko how much the taxi and museum tickets were and I was pleased that they were fairly inexpensive. Once we got to the British Museum, George surprised me and ordered the hologram guide, something he never does. So instead of using Yukiko to order a spoken guide from the museum, George had Yukiko order a hologram that spoke so charmingly and gave us a tour that Yukiko would know that we would want to have. It was such a thrill to see the British Museum just like Great-Grandma Samantha had so many years ago. I bet it has hardly changed.

We had such a big lunch that we made it to the early afternoon and went out for dinner. Yukiko found a nearby restaurant that we could walk to, and it had the food that Yukiko knows George and I like. The rain had stopped, so we decided to walk there, and it was a pleasant short walk. George and I always enjoy these walks. George has his Yukiko set to tell him about all the interesting historical things that he is seeing on the way. It does get a bit annoying to listen to details about where we are walking, but sometimes it was interesting to be reminded of how much this city must have suffered during the Second World War.

Lunch was very good. The conveyors brought us the food we ordered and the drinks were precisely what we had asked for. I was surprised at how much I liked the beer here, although maybe it was not a good thing to have with lunch when a person has a bit of jet lag.

We thought that we would go to another museum today but since the weather is better, we decided to go and walk around the city a bit. We rested just enough to have the energy to get through a long walk, seeing Big Ben, Parliament, and a few other sights. We were lucky that we could also get a quick tour of the Tower of London, since Yukiko notified me that they were open and there was availability. Although a bit jet lagged, we had a tour with Yukiko paying for the tickets and the rights to the Tower Tour. The tour was good and short enough, since Yukiko told them that we were tired from our trip and wanted a shorter one.

In the late afternoon, we went back to the hotel room. We forced ourselves to go and get an early dinner at a fish and chips restaurant, something making a comeback because of the advances in fish farming. The restaurant had the old-fashioned smell of fried foods, something I like. The robots brought us some tasty fish and chips. It was good that Yukiko recommended the restaurant that she did because the one George's Yukiko suggested was way too expensive, and I did not want to spend too much money on something that we would not be able to enjoy too much because of jet lag.

July 13, 2068

We woke up early and were ready for the day. Of course, we had a fulfilling breakfast at the hotel. It was a nice day and we wanted to see more of the city but Yukiko informed us that it was going to rain in an hour or so. We thought that this was useful information and took advantage of it. Since our hotel was close, we walked to the Natural History Museum. Great-Grandma Samantha did not have a chance to see it and I am glad that we did. George and I decided to just get Yukiko to pay for the official museum tour and Yukiko guided us around. It is an impressive museum.

Awful weather outside. After seeing this museum, Yukiko suggested taking a taxi to the Imperial War Museum that was close by. Yukiko ordered the self-driving taxi for us, and it came in just a few minutes. Although this is not really my thing, I agreed, and George loved the history there. Yukiko gave us the tour after paying for the entrance and the official information for the tour at the museum.

For fun, we took the Tube back to the hotel since the weather had improved. We went to Lambeth North Tube Station and Yukiko told us where to get off. There were so many people on the Tube, even if so many people work remotely these days. It is a busy city and it reminds me of my native New York City with the sounds of the delivery drones going overhead.

We had a great dinner at a restaurant where there were still people preparing food. Since humans were cooking the food, it was quite a bit more expensive than the robotic restaurants we usually go to but we thought it was appropriate, since this is the way that Great-Grandma Samantha experienced restaurants in her day. Who knows? There may be a time in the future when such restaurants are rare because fewer and fewer people get trained as professional chefs and people do not want to work as waiters.

July 14, 2068

We woke up on this Saturday feeling better and excited about what we would do next. My Yukiko knows that George is crazy about English history so she suggested to me a trip to Hastings, since the weather was good and it would be a fun day trip. I asked George and he told me his Yukiko suggested the same thing. George had his Yukiko get a tour to Hastings. The taxi took us to the train station and Yukiko paid for the tickets and told us which train to get on. We were in Hastings in less than an hour and Yukiko planned out the trip for us well. It was a pleasant day trip and we returned to London in the afternoon, a bit tired but glad that we saw something a bit different. I am sure that George loved seeing the sight of the battle that changed British history over a thousand years ago.

July 15, 2068

We got up in the morning and after breakfast went to the train station to get to Paris. We would be late for Bastille Day, just like great-grandma, but Bastille Day is much less popular than it was years ago. Now, Federal Europe Day is much bigger and that is in May. Yukiko was helpful but annoying. She ordered the tickets for us but when she noticed that we were eating breakfast a bit more leisurely than usual, she reminded us to get moving. The robots took the bags to the taxi and we were at the train station. It is good that we listened to Yukiko, since we made the train on time. The hyperloop was not in operation, since its tracks required some maintenance, so we took the old-fashioned train to Paris. Yukiko knew that the next train was full, so it would have been a bit of a wait to get the next train to Paris, so it is good we listened to her. Yukiko took care of the payments and we were scanned on the train once and at the border again. This time, there was no mishaps with George's facial recognition. I guess that the issue with his twin brother had been cleared up when we arrived in London, although we were ready for some issues with security. Sometimes when they use two redundant systems for security (facial recognition and microchips), we get issues with George's face. I guess the Europeans prefer the facial recognition technology over the microchip one, even if it is not as good as the microchip one. But that is a political issue, since so many people were against having microchips implanted into babies years ago in Europe.

We arrived in Paris in about two hours. It was nice to see the countryside from the train a bit.

Fortunately, Yukiko had booked a very convenient room in Paris, since someone had recently canceled their reservation and Yukiko picked up on it immediately. Yukiko got us a hotel at a great price because of her vigilance and we appreciate it. We got off the train in Paris and took the taxi Yukiko ordered and paid for to the hotel. There was a person who seemed to be the manager of the hotel. It seems that the computer system had gone down. Fortunately, Yukiko could translate in real time what the manager was saying. He assured us that our reservations were in order. Although we had to wait for about a half hour after the computer glitch had been fixed, the manager made sure that the reception processed our identities and payments immediately, when the computer system was running. We went to our rooms and the hotel's porter robot took our things to the room. While it was not a great view, the hotel was conveniently located and George was very happy about that.

We went out for our first French dinner in Paris. The robots cooked up some lovely meals and delivered all the courses well. George especially loved the cheese plates before the dessert. The robots recommended the smelliest cheeses to George and he just adored them.

July 16, 2068

It was a wonderful first full day in the City of Light and we were glad that the center of the city would have a huge dome over it a few years from now, since the climate has gotten a bit unpredictable and makes urban centers unpleasant at times. We walked the streets since Yukiko informed us that there was not going to be an issue with rain nor would it be oppressively hot during the day. So, we decided to walk around a bit and enjoy the city. There were many tourists and locals enjoying the wonderful weather and sitting with their friends in cafés around the city. We had seen outside the city where they are already preparing to set up the massive dome to protect the city. It will be fun to be in a city in which the temperature and rain are a bit more predictable and managed by the municipal authorities.

We had another amazing dinner; the robots here know how to make food well. And the wine is to die for with a wonderful fruity taste. It was an Alsatian restaurant, just like the one my great-grandma ate at. The smell of sauerkraut was in the air and though I usually prefer Asian food to European food, Yukiko planned this dinner well and it was a pleasant surprise, even though I am not a big fan of sauerkraut.

July 17, 2068

I had Yukiko set up some museums for us, this time I gave her some vague instructions and told her to surprise us. Yukiko saw that there was some availability at the Louvre in the early morning and alerted me to that. Immediately, she bought the tickets and rights to the Louvre tour. We took a self-driving taxi to the Louvre to get there in good time and Yukiko communicated to the 3D hologram what we wanted to see in the museum. We chose to have a 3D hologram tour instead of the virtual reality tour, since neither of us likes the headsets for virtual reality.

It was a long day and we enjoyed the tour, but by the early afternoon, we had had enough. Yukiko got us a taxi and recommended a restaurant.

After a good late lunch, we had a good walk in the streets, stopping once for a drink at a café and then went back to the hotel. We had a light dinner in our rooms printed by the 3D printer and went back to sleep. The 3D printers only produced a few basic sandwiches but it was enough for us this night.

July 18, 2068

One thing that I had heard about and wanted to see was Versailles, even if we have taken tours in Versailles using virtual reality back home. Yukiko knew that this was something that Great-Grandma Samantha saw so she built it into our itinerary. Yukiko reserved a tour in English at Versailles. While it would have been a bit cheaper to take the bus there, Yukiko thought it would be easier to just take a self-driving taxi. The taxi did not take long. It was a bit strange having a human leading the group in Versailles. She must be one of the only humans doing this still and she looked in her 80s. Because she did not seem to remember my preferences based upon what Yukiko informed her about, she still gave a good tour. I am not really used to having a general tour and not having a tour that is catered to my interests that Yukiko communicates to others. But it was good to see an old-fashioned type of tourist tour of Versailles and it was fun to think that this is probably how Great-Grandma Samantha experienced it. It was a bit more expensive but it was fun, I don't think there will be people doing this in the future, since the holograms and digital assistants may actually do a better job and are cheaper. Some of the older people still like to have humans as guides. Most of us younger people prefer more modern forms of guides.

July 19, 2068

We woke up late, since the walking and touring are taking quite a bit of energy out of us. It is good that we are going to the beach.

We took the latest train to Nice. Yukiko bought us train tickets from the Gare d'Austerlitz to Nice with the hyperloop. Once we were scanned in the train, we knew all we had to do was relax and sleep a little. We woke up at the station on the beautiful French Riviera. Yukiko had arranged for our hotel with a beautiful view. The taxi took us there and we were quite happy with it. Yukiko is sensing that we are getting more adventurous with food and is recommending slightly different foods every day. We even let her surprise us for dinner and she made her choices for us based upon our tastes. The cricket burgers were good, using a different sauce than the robots cook them with in the USA. Yukiko knows we have spent a bit more money than expected, I think she made the recommendations based upon our financial situation. Beef is really expensive and we ate it a bit too much in Paris. Crickets are cheaper to eat.

July 20, 2068

It is the Riviera and we are here to relax and enjoy the beach and sun. So that is what we did. We enjoyed a day at the beach breathing in the fresh salty air, since it was not too hot but was still warm and sunny.

July 21–24, 2068

We have really enjoyed the beach here in Nice. It is a pleasant place to be. But we have been exploring using local self-driving taxis here. We have used taxis to go and see Cannes and Cap d'Agde. It is nice that Yukiko is recommending various locations to me and I am sure she is making sure that my trip resembles Great-Grandma Samantha's trip so many years ago. I did ask Yukiko how much we are spending on taxis on the Riviera. It was a bit more than I expected and I have asked her to recommend some ways to save money, since I don't want to go overboard. She recommended taking buses to various beaches but George and I discussed this and would rather stay with the convenience of using the taxis, even if it is a bit more expensive, so we overrode Yukiko's suggestions

July 25, 2068

It is time for us to get to Rome for the return flight. I have gotten used to French food and will miss it. It did not seem so exciting but because Great-Grandma Samantha went to Milan, I thought I should too. Yukiko set up the trains for us and we took them to get to Milan. As usual, we were scanned but everything checked out, we could enter Italy and our tickets had been paid.

I liked Milan. I was really impressed with the cathedral and liked walking the streets there. But then, maybe I was just lucky because it was good weather and I was very relaxed from my time on the Riviera in France.

We spent a night in Milan and Yukiko suggested a good restaurant near the hotel. The food was good and it's good that Yukiko could explain what I was ordering because the names of some of the foods needed to be explained a bit. Yukiko is getting a bit interesting, even recommending that I try cow's tongue at the restaurant for dinner, something my great-grandmother was surprised by. I decided not to get that but it was something Yukiko is trying to do, to make me more adventurous with food, something I had asked her to do.

July 26, 2068

We took the train in the morning to Rome (less than two hours) and we really liked the city, even though it was very hot and humid. Building a dome over Rome to protect it and the inhabitants from the weather is in the plans, but they are not yet doing the prep for it, from what we saw. We actually met some people from Rome and had some discussions. They could not speak English and we cannot speak Italian but we had Yukiko translate. They suggested some places to see to Yukiko. Most of their suggestions were things that Yukiko thought that I would want to see anyway.

Anyway, Yukiko did it again and found us a great hotel and restaurant in a good part of the city at a good rate. From the smell of it, the mopping robots seem to use the same cleaner as the ones in London. I am so glad that the hotels do not force us to share a bathroom down the hall like so many of them did in great-grandma's days.

July 27, 2068

Walking the city today has been fun. Since George is so crazy about history, this is the city for him. Yukiko was telling us the background information of so many of the things we were looking at. George and I learned a lot about the history of the city. Yukiko has also begun to teach us some Italian during our walks on the trip, as we had instructed her to do.

July 28–29, 2068

We wanted to see all we could in Rome so we stayed until we had to return on Monday. Yukiko found us a great hotel and we wanted to enjoy this time in this great location. Every night Yukiko recommends a better restaurant and every night she gets me to eat more adventurous foods, as I always wanted to do. I am starting to learn some Italian, since Yukiko insists on teaching us words and phrases. As in Paris, Yukiko brought us to good museums and we have enjoyed seeing and learning from the museums we have visited.

July 30, 2068

Unfortunately, every holiday has to end. Yukiko made sure that we were checked in and ordered the taxi to the airport. A human driver! They are so rare now because most taxi cars do not even have steering wheels. The chat with the driver was pleasant but the constant gesturing with his right hand (and sometimes with both hands!) during our conversation made me feel uneasy as I thought that any moment we would go off the road.

When we arrived at the airport and were scanned, the porters brought our bags to the check-in, and we went through security, scanning our belongings, biometrics, and microchips. It was a pleasant holiday, and I look forward to my next one. Yukiko did surprise us a few times with restaurants and ordering us food, and since she knows us both so well, it was mostly successful.

George is going back to work in two days. The holiday was an adventure for us and we will treasure it forever. Maybe my great-grandchild will find this file and do what I have done. But then, maybe Yukiko will just tell her about this trip.

0.5 The Transformation from 1968 to 2068

Craig Webster

How the World Has Changed and How It Will Be

The year 1968 was a memorable year in the modern world. Many readers do not think of the people in that period as living, thinking and working radically differently from today. In a political sense, so much happened that year that was memorable and has shaped our modern world. For example, protests all over the globe changed so much, both politically and socially (Barone *et al.*, 2022; Carey & Andrea, 2016; Maeckelbergh, 2011).

The student protests in Paris almost destroyed the entire political regime of France. The protests in Prague that led to the invasion of Czechoslovakia by several Warsaw Pact countries indicated to the world the extent that the Soviets would allow for experimentation with socialism. Robert Kennedy and Martin Luther King were assassinated in that year. Specifically, the assassination of Robert Kennedy led to a very bloody and violent Democratic National Convention punctuated by street violence in Chicago, when such political spectacles are usually exemplified in the United States by speeches, music, balloons and confetti. These protests were not limited to a few countries, as massive and meaningful protests and political events made 1968 a hugely important year for many countries.

While 1968 was a volatile year for many countries, much more volatile than one would have expected, the other major international factor that differs very much from the era we now live in was the Cold War and the notion that the world was split between two major superpowers and their allies, with the remaining countries belonging to a 'Third World'. What was noteworthy about this is that the Vietnam War became very contentious in the United States and elsewhere, as the Tet Offensive of 1968 illustrated that the Vietnamese forces that sided with the communists were organised and capable of disrupting the entire country, seriously threatening the United States and its allies. While the Tet Offensive was eventually a victory for the United States and its allies, the communist offensive illustrated that it was possible to attack US forces in a sustainable way, so that many Americans began to think that the Vietnam War would ultimately be lost.

From the political world in which people were living in 1968 to the transformation of the world of 2068, we see a number of changes in terms of technologies that have made major differences, some of which are more commonplace and understandable than others. There are major technological changes in terms of transportation, finance, security, logistics, translation and others that have made and will continue to make the experience different and likely more pleasant into the future. From the perspective of the tourist, the experiences of 1968 and 2068 will be very different, since nearly all aspects of the tourism experience will be transformed. Here, we provide an overview of the transformation of tourism from 1968 to 2068, from the perspective of the tourist, making explicit references to how automation technologies will have advanced and solved many of the challenges that faced tourists in 1968. In addition, many back-of-house operations, which are of no concern to most tourists, will have been automated and made more efficient.

Transportation

1968 to today

While 1968 seems like a modern period for many, transportation was a bit different from what many travellers now experience. To begin with, air flight had advanced to a point where flights on jet aircraft were no longer a novelty. Boeing 707 jets were able to transport people quickly and relatively cheaply and were widely in operation by the late 1950s. But there were qualitative differences in the nature of international flights, since jets were louder for passengers, smaller and less fuel efficient than the Boeing 747 and some other larger jets that now carry large numbers of passengers. In 1968, international jet travel was happening but it was less common, more expensive and, in some ways, less convenient and pleasant than it is now. For example, the fuel efficiency problem meant that transatlantic flights had to make stops in places such as Newfoundland, Iceland and Ireland to refuel. While there was more legroom, better service and better food, flights were slower and louder, they had to stop to refuel and there were fewer flights from which to choose. In our current times, air rage is an increasing problem (McLinton *et al.*, 2020) likely because of the lack of space and many other stresses placed on air travellers.

Apart from air travel, rail travel was also somewhat different in 1968 than it is today. For one thing, all the connections that we now have that enable tunnels and bridges to connect islands were not as developed as they are today. Beginning in the 1930s, a person could take a night train from London to Paris (the 'night ferry'); however, it was not as seamless a trip as people have today. For example, the sleeping carriages would have to be loaded and unloaded from the ferries at the ports, slowing down the transportation time and making the entire trip about 11 hours. The night train from London to Paris was not alone. Some infrastructural issues that have only been addressed in recent years, such as the building of a bridge between Sweden and Denmark and the bridging of various islands in Denmark, have resulted in much more seamless rail transport in these countries. The increased construction of bridges and tunnels to assist the logistical process of moving people and cargo has done much to quicken travel with the best-known example being the Channel Tunnel connecting the UK and France.

Rail travel is not just an issue of seamlessness, but it is also a question of having the infrastructure and the technologies to have faster trains. In 1968 tsleeper cars in trains made a great deal of sense, since trains were slower and often did not have the seamlessness of current or future train travel. , The lessening of the physical and political impediments to travel along with the increased technological ability to transport people quickly on the surface is an advantage of train travel. The major issue with regard to this in terms of the future is whether rail travel is seen as environmentally beneficial, fast, cost-effective and convenient.

Buses and automobiles have largely the same strengths and weakness. Regarding buses, in 1968 and up to today, they have, to a great extent, the same advantages over other forms of transportation. Buses and automobiles can both drive where roads are currently in place. So it may mean that these forms of transportation can bring people to destinations that have workable roads. However, what differentiates the bus from the automobile is that buses require a paid driver and run to a schedule that may or may not be convenient for a traveller. Buses have the disadvantage of being developed to a schedule and lack the flexibility that an individual traveller may wish for. In terms of current buses and automobiles, a driver is needed. So, in this way, a sober, attentive and licensed driver has to be available to take control of the bus or automobile.

2068

In 2068, we will not be living in a world with teleportation, although there is some theoretical work to enable the technology to function (see, e.g. Li *et al.*, 2021). Noteworthy developments in other forms of transportation will be implemented in the next few decades. It is unlikely that some technologies such as gliders or air balloons will be used for large-scale tourism transportation, although they will likely continue to be used for leisure activities and exploited for such in the tourism industry. However, improvements in air, rail and automobile technologies will change a great deal in terms of the cost, travel distance, speed and comfort by which people travel.

While there will undoubtedly be improvements in terms of building infrastructure such as bridges and tunnels to connect various islands and land masses that are separated by water, this is not the primary source for improvements in travel. To begin, we see that there is much focus on fuel efficiency, since fuel is a cost and is linked with the sustainability of travel. As such, there is a great deal of political focus on rail and bus travel. Since these forms of transport are associated with fuel efficiency, there will be further investment in them for the foreseeable future. Indeed, the Channel Tunnel (opened in 1994) alone did much to modernise Europe's infrastructure and made significant contributions towards sustainability in transport (Goldsmith & Boeuf, 2019).

Another innovation that will be widely used by that time is autonomous vehicles (Dimitrakopoulos *et al.*, 2021), since automation technologies will make it so that humans will not be required to steer and pilot vehicles, whether in the air, on the sea or on land. One obvious change that this will bring about in the economy is that employment in logistics will be entirely different, since one very common profession, the truck driver, will be eliminated with this innovation. However, it is expected that this technology will make it easier and safer for individuals to travel in automobiles. The practicality of individual cars will remain attractive to many because of convenience, and the self-driving feature will make the automobile a more social environment. It is likely that governments will provide incentives to discourage individual automobiles and encourage public transportation. Alternatives to this will include ride-sharing arrangements in which citizens will become members of a sharing group and access automobiles in ways that are optimised using artificial intelligence (AI) predictive software.

Financial Technologies

1968 to today

One of the realities of life is the need to have money to pay for goods and services. In 1968, many interactions for daily needs used physical forms of money, such as cash or check. While credit cards certainly existed, they were not widely used, since the mechanisms for processing them required a lot of manual manipulation and bookkeeping. An issue that travellers in 1968 had to contend with was paying for things using cash in the local currency. The financial world in 1968 was different from today's financial world. There were different currencies, and currencies had to be used in physical form, for the most part. Visitors to a country would often bring a different form of money, the traveller's check, and convert the checks either at a tourism establishment or at a bank. An alternative opportunity for a traveller was to bring large sums of cash, something that many visitors would be unlikely to want to do for security reasons.

So, visitors in 1968 had to contend with different forms of currency that may not have been easy to understand or convert into their home country's currency. In addition, the physical requirement of changing cash or traveller's checks would mean that a tourist would have to have the resources at hand and be at an establishment willing to accept either traveller's checks or another currency or go to a bank or other institution that would change traveller's checks. Banks across much of the world do not have opening times designed for the customers' convenience, often being closed during weekends, holidays and at night. In 1968, institutions at airports and train stations would change traveller's checks for the local currency, for a commission. In addition, travellers had two problems, one of which is historical and one that has largely been solved.

The first problem a traveller would face in 1968 was estimating how much money to change to use in a country. This would require estimating expenses in a currency with costs that were not entirely transparent. What this would mean is that if a person estimated badly, they could have a problem finding a way to get through a weekend, in the event that no institution was willing to change cash or traveller's checks.

The second problem, admittedly a minor one, is that a tourist would likely accumulate a handful of coins while travelling. Since modern economies still use physical currency and many transactions require coins, tourists frequently accumulate small (or large) amounts of foreign coins that are not easily used outside of the country where a particular currency is used. There are likely small amounts of Italian lira, Deutsch mark, French francs and other antiquated coins in many homes from a time when people travelled and accumulated these coins. With the advent of the euro, many travelling to Europe will have fewer of these metallic souvenirs. In the future, a common currency may replace the Canadian, Mexican and US currencies. Additionally, the increasing use of cryptocurrencies that are only digital may mean that this minor issue will be largely historical. However, currencies based on blockchain technologies have some issues regarding sustainability (Giungato et al., 2017) and if this issue is not ameliorated, cryptocurrencies based on blockchain technologies may not have a very robust trajectory.

2068

Financial innovations have changed and will change many things. While financial systems are imperfect, we will have almost entirely abandoned all physical forms of currency. It would be a mistake, though, to imagine that people will still carry credit cards. Currently, most credit cards use a radio-frequency identification (RFID) chip to identify the card. It is likely that future technology will not be something that is separate from the individual. In the future, the RFID chip or a similar technology will be embedded in the human body (Ivanov et al., 2014), so that the individual will not be easily separated from the identifying chip, a chip that will be linked with various forms of personal and financial information. In addition, other biometric information will be used to identify individuals with their payment method – e.g. the use of face recognition technology or fingerprint readers to link a person to their bank account (Nasution et al., 2020).

There will be risks with embedded financial and security technologies. The risks are not just the questions raised about what such technologies pose to human health and well-being but also have a great deal to do with how they can be used by unethical actors. For example, if an RFID chip is embedded in a person's hand, it is possible to spoof the vibrational field of the chip and thus hijack a person's wealth. However, in a more low-tech solution, a thief could simply cut the RFID chip out of someone and use that chip to pay for goods and services for the thief's benefit. So, while such embedded technologies may have some benefits, e.g. nudists can pay for drinks at a beach bar without having to have a credit or debit card in hand, the risks of high-tech or low-tech theft of the wealth linked with a chip are real. Future technologies will increasingly link an individual's biological information with the embedded technology so that the technology will not allow use of the financial device when separated from the person authorised to use it. Already, there has been a foray into the use of redundant technologies to ensure the legitimate use of funds (see, e.g. Karovaliya et al., 2015).

Political and Security Technologies

1968 to today

In 1968, borders existed between countries, as they still do. The presentation of passports at many borders is expected and accepted by most people. There are some exceptions to this, since there are few barriers to the movement of tourists in the Schengen Area.

While the expectation of tourists was that their passport would be inspected by a human at a border in 1968, massive technological developments have allowed for the automation of this process. For example, while human agents are still used to record the entry of people into many countries vetting incoming tourists, to some extent, technologies have already automated a great deal of the process, so that it becomes easier for the human agents who remain. Already, many countries allow foreigners to enter by scanning their own passports and having facial recognition technology verify the matchup of the passport with the incoming person. So, while human agents are still the general rule to ensure that the person who holds the passport and the passport are genuine and belong together, some security technologies have been integrated into the system allowing for a more automated process. The automation technologies used today exhibit some imperfections and will need improvement before they can be used with little human oversight into the future (Khan & Efthymiou, 2021).

2068

In 2068, many humans involved with border security will have been eliminated. This does not necessarily mean that there will be human-looking robots or holograms equipped with AI to interrogate individuals at the border and to ensure that their paperwork is in order. More likely, future border security will be automated using descendants of the current facial recognition technologies and will likely be strengthened with embedded technologies that will be descendants of RFID chips. While the physical passport may remain in use for some years, some form of embedded technology that will be tightly linked with the individual will be used more frequently (Ivanov *et al.*, 2014). Eventually, if the health risks do not seem to be unacceptable, an embedded technology will be used to identify individuals. Such a technology will require political cooperation between countries to ensure that individuals can be tracked from one country to another. While any embedded technology can be hacked and spoofed, a secondary technological check will be carried out to ensure that the person with the embedded technology is the person who is officially allowed to have that embedded technology. That technology will likely be a facial recognition technology or another technology that uses biometric data from an individual.

While these technologies to identify individuals will be used for crossing borders, they will also be used to track individuals and ensure security once a person is at a destination. Such an embedded technology, perhaps also using biometric data such as facial recognition, will also provide security in hotels and other venues in which a person is permitted to have access. So, while the technology used in 2068 will largely be developed to manage the movement of people between countries, it will also be used to ensure identities and security in private spaces. This sense of security will also be important with regard to the use of financial instruments, many of which will have an embedded element, since physical forms of money will largely be phased out.

Technologies to Porter

1968 to today

In 1968, tourists carried their own luggage, for the most part. While, occasionally, porters were used to carry bags to a room and transport bags from a taxi to a hotel, for the most part, tourists carried their own bags. One remarkable aspect of this is that in modern travel, many tourists have suitcases with embedded wheels. The idea of the wheel being incorporated into suitcases is a very recent invention and something that the people in 1968 would not have had but would have appreciated. Additional portering technologies have been developed such as the moving sidewalk found in many airports that transports people and their possessions from one place to another, minimising human exertion.

2068

The technologies to carry things for tourists will be very helpful to humans in 2068. In any location where humans need to carry things, robotic technologies linked with AI will lighten their burden. Robots will pick up, transport and deliver baggage when needed. It is unclear whether travellers will use technologies at various venues or bring their own to porter belongings. An additional question with regard to such robots is if hotels, restaurants and other venues own the robots or outsource the robotics used in their facilities (Ivanov & Webster, 2020). It is conceivable that a few robotic manufacturers will dominate the industry, but manufacturers choose not to be overly involved in the daily applications of the robots and stick to the business of simply manufacturing the technology and dealing with service issues. However, it is also conceivable that there could be the development of massive companies that would own, rent and lease such robots to individuals and companies in the field and be separate businesses. While it is still likely that robots that porter baggage will be widely used, especially in transporting a tourist's belongings in a hotel, the business model that will predominate is still unclear.

Language Technologies

1968 to today

In 1968, language was an issue because not all public education systems put the same value on or had the same ability to teach foreign languages to students. As a result of educational systems' choices and capacities and the reality that not every person had the ability to learn languages or an interest in learning foreign languages, language was a problem for the tourist. The problem was dealt with using the means available. Those organisations that catered to tourists would hire personnel who could speak foreign languages, when such people were available. In addition, tourism establishments had signs that were understandable without using language, such as signs indicating bathrooms for males or females. And some organisations had the ability to translate tourism information into a few chosen languages that a tourist would likely speak.

As such, tourists would often find themselves in a situation in which language was a serious problem. They not only may not have been able to understand what was written on a menu but they also may have had a hard time communicating with the waiting staff who may not have been able to translate some key words on the menu. The problem of language and comprehension of others was not something that a person faced on occasion and perhaps only at a restaurant or a museum but was a constant issue with which a person would have to contend. For example, a train conductor may not have been able to understand a tourist's questions about the next stations or other directions and this could make a simple thing such as knowing which train station to leave a train difficult for a person to discern. Some tourists would have phrase books to translate common phrases, but this was always a clunky and inadequate tool to bridge the language gap.

English is thought by many in the West to be a lingua franca in the present day and some may believe that it will always be possible to navigate situations using English. This is clearly not true, since not all tourists are conversational in English and not all who encounter tourists (passersby on the street, museum guards, ticket agents and others) can communicate well in English. While in 1968, people made the best of the situation by using maps and guidebooks, and would seek those who spoke a common language with them, a traveller in the current era has access to the internet and other technologies that allow a person to bypass the need to ask individual humans for directions or insight into various aspects of the destination visited.

2068

In 2068, the language barrier will have been almost entirely eliminated. While current technology is used to assist in language learning (Jiménez-Crespo, 2017), future technologies will largely bypass the human learning process. Translation technologies for spoken and written language will have been perfected, meaning that communication with people or the interpretation of signs and other written materials will be easily translated with the technologies available. In 2068, a person will be able to speak only one language and be able to communicate with many others who only speak one language, since translation will be done in real time and will be highly accurate.

While the development of technologies alone that can communicate in multiple languages is important, their implementation into tourism establishments will be most appreciated. It is conceivable that, in 2068, tourists will be able to understand information from multiple sources in the language requested. For example, while a few people may use telephone-like devices to get tourist information about what is available in a museum, kiosks in the museum may provide such information. In addition, robotic guides in museums may also provide guided tours that would be tailored to the wants and needs of a specific visitor and in the native language of the visitor.

As such, the language barrier will not only be minimised, but the technologies for language translation will be integrated into much of the robotic technologies that will be used in all facets of the tourism industry. That is not to say that there will not be limitations. The dialects of some languages may be challenging for translation to handle, although many dialects may be extinct by that time (Zhang & Mace, 2021), thereby minimising this hurdle.

The Integration of Hardware and Software into the Tourism Ecosystem

1968 to today

In 1968, the hotel, the restaurant, the museum, the train station, the airport and almost any other physical space that a tourist would use was not nearly as automated as the analog of today, let alone the analog of 2068. The tourism experience in 1968 was often highlighted by human contact and human labour. In 1968, while there were guidebooks and signage, tourists would ask for directions and information from people. In addition, many of the services were provided by humans. For example, in 1968, if a person wanted to cash a traveller's check or change money, that person would have to find a person working in an establishment to do such a transaction. For the modern tourist, such an interaction with a human would likely be unnecessary using current technologies, since many transactions may not require cash, and cash in the local currency can often be withdrawn from an ATM. A modern tourist, too, may not have to ask for local information and directions, since global positioning systems (GPS) and internet technologies provide the tourist with a great deal of independence from relying on information from locals at a destination.

The **hotel** in 1968 relied on human staff who would staff the reception as well as do all the other manual work that needed to be done. Reception, alone, requires not just emotional intelligence to deal with customer service issues but also communication skills. In 1968, the ability of the reception to communicate in any language other than the local language was not a given. A human working at the reception would (and still does) face language issues with guests. So, the tourist of 1968 would face the difficulty of finding a hotel and sometimes struggling to communicate information regarding room availability, rates and other relevant information. Another issue faced by travellers in 1968 was finding an available room. Many times, visitors would have to call or walk from hotel to hotel to find available rooms. Hotel reservation services were available in some places to assist with this, but many times locating available rooms was a hit or miss issue.

Much of the accommodation of 1968 was different from today's or the future's accommodation. Progress has largely moved many accommodations away from shared bathrooms and shower facilities. In addition, the accommodations did not offer the entertainment of the modern hotel. Technology in terms of entertainment (television, cable, internet and streaming services) has advanced and changed some aspects of the hotel guest's experience. However, for the most part, the experience is largely the same, even with room service generally being served by humans.

The **restaurant** of 1968 required a great deal of human labour, with cooking, cleaning, serving and billing all done by humans. For visitors, especially international ones, language concerns were a major issue, since understanding a menu in a different language was not always easy. Some restaurants in heavily touristed areas would have menus available in other languages but this may not have been entirely helpful, since such menu translations were limited. Because of a reliance on human labour, restaurant staff may not have been able to translate menus for tourists. In addition, menus were less globalised and less sensitive to the special needs of particular tourists, so that a tourist who only ate vegan, halal or had food allergies may not have been catered for by the restaurant.

The **museum** experience of 1968 required a great deal of information gathering such as knowing the opening hours, the names and locations of museums, as well as information on current exhibits. As such, reliance on guidebooks, hearsay from previous visitors, assistance from an accommodation's concierge/front desk and tourism authority official materials would be vital. Modern technologies have made this a largely easier exercise in gathering information prior to a museum visit, since much of the information needed is available from internet sources and can be available in multiple languages. In the museum of 1968, little information was made available to foreign tourists since museums were largely reflections of the goals and values of the government, whether it be to foster nationalism or promote socialist principles to the visitor (for some museums to the east of the Iron Curtain). In recent years, this has also been ameliorated with curation in other languages with headphones and digital technologies, even if the cultural values may be foreign to the visitor.

Transportation experiences (airports, bus stations, train stations and taxis) are a big part of the entire tourism experience. In 1968, transportation required gathering information by reading and interpreting material, much of which was in a foreign language. So, even interpreting the signage in transportation hubs could be a challenge. Gathering information, understanding the best ways to travel, methods for purchasing tickets and the speed of travel made the experience much more difficult than the current process.

2068

The **hotel** of 2068 will be radically different from the hotel of 1968, since it will be much less dependent on human labour, and digital assistants may share information and make decisions to make the stay more fluent. In the hotel, there will be a reduction in the use of human labour, since automation technologies will make humans largely superfluous. Even before they arrive at the hotel, guests will have used AI to find, book and pay for a room, assisted by the visitor's digital assistant. AI will not require a robot that resembles a human being but may simply be activated by voice. Robotic technologies will carry baggage from the vehicle upon the guest's arrival. The check-in process will be automated and may not even require a front desk because with embedded technologies to identify the guest, they would be logged, allowing the guest access to the hotel as well as the particular room booked. While in the room, all the necessities of a hotel visit will be automated, most likely with voice-activated commands for the particular needs of the guest. Many of the services may be fully automated, such as room service, where a guest could order a meal and have it sent to the room using a service robot and have the bill attached automatically to the visitor's account. One benefit that the 2068 tourist will have is that there will be little or no issue with communication, since the software will be able to recognise and communicate in many different languages, and digital assistants will share relevant information to improve the experience for the visitor. To some extent, the visitor will also be different. Many of them will have more time, although they will likely be less wealthy than middle-class Westerners are today and will, therefore, need their digital assistants to find the most efficient and cost-effective experience. There may also be new conceptualisations of hotels, putting some hotel rooms on wheels, similar to automated caravans, of sorts.

The **restaurant** of 2068 will be somewhat different from what is now experienced. Restaurant patrons will be able to better understand what foods they are ordering and will be able to communicate with AI to ensure that they do not order foods that would conflict with their tastes/religious restrictions/allergies. Food will be prepared and delivered by robots after receiving the order from the software system. While there will likely still be tables where restaurant patrons will eat, delivery and cleaning systems will be automated. All these systems will be integrated, so that information from the restaurant patron's order will be passed to the robot. In addition, payment will be automated, likely using the embedded technology of the restaurant patron. There will also likely be establishments with human chefs and servers, but these will largely be available to wealthier patrons. In addition, some food service may be provided by 3D printing technologies (Nachal *et al.*, 2019). In addition, augmented reality (AR) technologies will be increasingly integrated into the restaurant experience (Ali, 2022).

The **museum** of 2068 will be somewhat similar to that of today but there will be more automation, in many respects. As with many other aspects of the tourism system, much of the information and tickets will be automated and virtual, and security and payment will take advantage of biometric information and embedded technologies. While patrons of museums may stroll the hallways and enjoy the artwork and historical artefacts, many may choose to have a more curated experience with a guide that may be either robotic and interact with them or there may also be the opportunity for holographs to give tours. Virtual reality, AR and holograms will add to the visitors' experience. These technologies may allow for a more immersive experience, enabling visitors to experience historical eras and integrate themselves into an edutainment experience (Serravalle *et al.*, 2019) in their own language.

The **transportation** ecosystem (airports, bus stations, train stations and taxis) of 2068 will be highlighted by easy access to information in the traveller's native language, autonomous vehicles and automatic payment systems. In future transportation, as in many other elements of the tourist's ecosystem, there will be few problems with language and communication. The software to communicate in an effective way with a person in their native language will streamline the gathering and acting on information as a process. In addition, AI will ensure that the most helpful information is brought to the attention of the traveller. Apart from that, autonomous vehicles will change how transportation takes place, since a human operator will not be needed. While, currently, little communication is needed between those who pilot vehicles in the air and on buses, ships and trains, more communication is needed for other forms of transportation (taxis and other automobiles). The nature and logic of the autonomous automobile, as opposed to the other forms of transportation, will require a redesign, since taxi and automobile rides will be a much more social experience than it is now (Cohen & Hopkins, 2019). As with other components of the tourism industry, payment will largely be made via the embedded technology of the user.

Automation from 1968 to 2068

The great transformation in this century for tourism was the movement from a less-automated version of a tourism ecosystem to a capital-intensive tourism ecosystem. The transformation has been driven by two major factors – the shortage/shortcomings of human labour in developed countries and the technical abilities of new technologies. The demographic factor is a major one, allowing for a shrinking labour pool in developed countries to create a situation in which the replacement of human labour is a necessity (Webster & Ivanov, 2020). While there are other means to increase the available labour supply (encouraging fertility and immigration), each of these approaches seems to suffer from weaknesses such as the general inability of governments to encourage people to have more children and the problems of integrating new populations into a host country. The fact that technologies have moved from conceptual to practical during this period, too, is a major factor. While in 1968 there were jet airplanes, ships, helicopters, trains and automobiles, the thought of them being autonomous and not needing a human to pilot them was merely science fiction. Now, such technologies are a reality, as with so many of the other technologies that will transform the tourism experience.

While here we have largely dealt with what tourists see and experience, there are back-of-house changes that they would not experience. Many of the dangerous, dull and dirty jobs will have been automated by 2068. Thus, many of the things that most tourists never even have to think about or see (such as maintenance of the properties, garbage removal, storage, human resource management and food preparation) will have been automated.

Two of the difficulties many travellers had in 1968 had to do with language and paying for things. Modern technologies that are now widely used have largely solved these difficulties. Travellers may use the internet and computers to assist in translation and generally have few problems paying for things using debit/credit cards linked to the global financial system. While the current translation technologies are imperfect and may not be as sophisticated as the next generation, future technologies will certainly minimise language divides further. Additionally, the embedding of further sophisticated security technologies will make payments a far more seamless and cheaper process than current-day tourists experience.

There are questions about how the technologies will be implemented in a practical way. For example, while the setup of a restaurant currently involves tables and booths to which food is delivered, a reimagining of the setup of the restaurant may happen. For example, current technologies make the conveyor belt system for delivering food, as is popular in sushi restaurants, a technologically feasible and effective system. The restaurant will likely require a very significant redesign of the kitchen and storage areas, as these are where deliveries, the storage of ingredients and the preparation of foods will take place, but using little human labour.

Hotels will also be redesigned in ways to enable automation technologies to be most effective. Hotels will have to keep in mind that many robots used in the facilities will have very different needs from human workers. Robots will need places to be stored, repaired and charged. Robots will also require a robot-inclusive design of hotel facilities that allows them to move and function around the hotel (Ivanov & Webster, 2017).

At this point in history, it is hard for us to imagine going backwards, in a technological sense. While there may be significant turbulence politically, culturally and economically in the world in the upcoming decades, it seems that whatever technological advances have been made will neither be forgotten nor go unused. It seems that we are entering into a time in history in which automation technologies will enable humans to step back and work less, since there will be fewer tasks over time that only human labour can do. Only a major calamity, something that would be difficult for most of us to foresee, could result in humans returning to a darker age with less technological savvy. For tourists, this means that automation technologies will be at their service, even if there are imperfections and problems with the ways that each of these technologies are employed in the practical setting of travel, tourism and hospitality.

Our grandchildren, we surmise, will likely have a highly automated vacation, although there will be a market that is wealthy enough to pay for a digital detox experience that uses more human employees. For the most part, future tourists will not have to struggle to communicate with others nor search in vain for basic travel information. With ease of travel, they will have a less stressful experience, although one has to wonder if part of the adventure of travel will be lost, as some of the charm of travel is the accidental discoveries and meetings that lead to long time memories and relationships. Doubtless, future tourism will discover how the yearning for adventure can use new automation technologies to make satisfying and interesting travel for all.

May the Robotic Force Be with You!

References

Ali, F. (2022) Augmented reality enhanced experiences in restaurants: Scale development and validation. *International Journal of Hospitality Management* 102, 103180. https://doi.org/10.1016/j.ijhm.2022.103180.

Barone, G., de Blasio, G. and Poy, S. (2022) The legacy of 1968 student protests on political preferences. *Economics Letters* 210. https://doi.org/10.1016/j.econlet.2021.110198.

Carey, E. and Andrea, A.J. (eds) (2016) *Protests in the Streets: 1968 across the Globe*. Hackett Publishing.

Cohen, S.A. and Hopkins, D. (2019) Autonomous vehicles and the future of urban tourism. *Annals of Tourism Research* 74, 33–42. https://doi.org/10.1016/j.annals.2018.10.009.

Dimitrakopoulos, G., Tsakanikas, A. and Panagiotopoulos, E. (2021) *Autonomous Vehicles: Technologies, Regulations, and Societal Impacts*. Elsevier.

Goldsmith, H. and Boeuf, P. (2019) Digging beneath the iron triangle: The Chunnel with 2020 hindsight. *Journal of Mega Infrastructure & Sustainable Development* 1 (1), 79–93. doi: 10.1080/24724718.2019.1597407

Giungato, P., Ranam, R., Tarabella, A. and Tricase, C. (2017) Current trends in sustainability of bitcoins and related blockchain technology. *Sustainability* 9 (12), 2214. https://doi.org/10.3390/su9122214

Ivanov, S. and Webster, C. (2017) Designing robot-friendly hospitality facilities. *Proceedings of the Scientific Conference 'Tourism. Innovations. Strategies'*, 13–14 October, Bourgas, Bulgaria, pp. 74–81.

Ivanov, S. and Webster, C. (2020) Robots in tourism: A research agenda for tourism economics. *Tourism Economics* 26 (7), 1065–1085. https://doi.org/10.1177/1354816619879583

Ivanov, S., Webster, C. and Mladenovic, A. (2014) The microchipped tourist: Implications for European tourism. In A. Postma, J. Oskam and I. Yeoman (eds) *The Future of European Tourism* (pp. 86–106). Stenden University of Applied Sciences.

Jiménez-Crespo, M.A. (2017) The role of translation technologies in Spanish language learning. *Journal of Spanish Language Teaching* 4 (2), 181–193. https://doi.org/10.1080/23247797.2017.1408949

Karovaliya, M., Karedia, S., Oza, S. and Kalbande, D.R. (2015) Enhanced security for ATM machine with OTP and facial recognition features. *Procedia Computer Science* 45, 390–396. https://doi.org/10.1016/j.procs.2015.03.166.

Khan, N. and Efthymiou, M. (2021) The use of biometric technology at airports: The case of customs and border protection (CBP). *International Journal of Information Management Data Insights* 1 (2), 100049. https://doi.org/10.1016/j.jjimei.2021.100049.

Li, J., Fang, X., Zhang, T., Tabia, G.N.M., Lu, H. and Liang, Y. (2021) Activating hidden teleportation power: Theory and experiment. *Physical Review Research* 3 (2), 023045. https://doi.org/10.1103/PhysRevResearch.3.023045.

Maeckelbergh, M. (2011) The road to democracy: The political legacy of '1968'. *International Review of Social History* 56 (2), 301–332. https://doi.org/10.1017/S0020859011000162

McLinton, S.S., Drury, D., Masocha, S., Savelsberg, H., Martin, L. and Lushington, K. (2020) 'Air rage': A systematic review of research on disruptive airline passenger behavior 1985–2020. *Journal of Airline and Airport Management* 10 (1), 31–49. https://doi.org/10.3926/jairm.156

Nachal, N., Moses, J.A., Karthik, P. and Anandharamakrishnan, C. (2019) Applications of 3D printing in food processing. *Food Engineering Reviews* 11, 123–141. https://doi.org/10.1007/s12393-019-09199-8.

Nasution, M.I.P., Nurbaiti, N., Nurlaila, N., Rahma, T.I.F. and Kamilah, K. (2020) Face recognition login authentication for digital payment solution at COVID-19 pandemic. In *Proceedings of the 2020 3rd International Conference on Computer and Informatics Engineering (IC2IE)*, 15–16 September, Yogyakarta, Indonesia (pp. 48–51). IEEE. 10.1109/IC2IE50715.2020.9274654

Serravalle, F., Ferraris, A., Vrontis, D., Thrassou, A. and Christofi, M. (2019) Augmented reality in the tourism industry: A multi-stakeholder analysis of museums. *Tourism Management Perspectives* 32. https://doi.org/10.1016/j.tmp.2019.07.002.

Webster, C. and Ivanov, S. (2020) Demographic change as a driver for tourism automation. *Journal of Tourism Futures* 6 (3), 263–270. https://doi.org/10.1108/JTF-10-2019-0109

Zhang, H. and Mace, R. (2021) Cultural extinction in evolutionary perspective. *Evolutionary Human Sciences* 3, E30. https://doi.org/10.1017/ehs.2021.25

Part 2
Foundations of Robonomics

1 Principles and Drivers of Robonomics

Stanislav Ivanov

1.1 Introduction

The concept of an automated economic system is not new. For over two and a half centuries, companies have used various technologies to increase their productivity, expand their markets, decrease production costs per unit and be more competitive (Frey, 2019). Societies have experienced waves of technological changes that have led to greater levels of automated production and lower levels of human involvement in it. The invention of the steam engine, the spinning jenny, the power loom and trains fuelled the First Industrial Revolution at the end of the 18th and the beginning of the 19th century. It was associated with the increased production of coal, textiles and other products, improved factory design and operations, and the start of the mass production of consumer goods. However, most of the benefits of automation were reaped by industrialists. Technology was used to replace workers and, logically, the latter resisted it. The Luddite movement in Britain from 1811 to 1816 is a good example of this direction. The Second Industrial Revolution, which started around 1870 and involved much of the first half of the 20th century, saw invention of the internal combustion engine and cars and communication technologies such as the telephone and the telegraph, advances in chemistry, electrification and steel production and more. Introduced by Ford in 1912, the production line epitomises the mass production that brought the fruits of automation to people's homes (electricity, cars, telephones, home appliances, medicines), while Chaplin's *Modern Times* presents the dark side of that same manufacturing process. The Third Industrial Revolution began after the Second World War and was driven by nuclear power, the development of computers, the internet and the digitalisation of production. Nuclear power provided humanity with a significant energy source while computers and the internet digitised and transformed organisational processes and workplaces and created new business models.

As a whole, the industrial revolutions were extremely beneficial to humans in terms of well-being, quality of life, longevity and health.

Life expectancy in 1800 was around 34 years in Europe and 35 in the Americas, but only 26 in Africa and 28 in Asia. Over two centuries, these numbers have more than doubled. In 2015, a baby born in Europe could expect to live on average 78 years, 80 years in North America, 75 years in South America, 72 years in Asia and 61 years in Africa. In many countries, life expectancy is over 80 years (Roser et al., 2019a). At the same time, child mortality has dropped sharply – in 1800, 43.3% of children did not survive the first 5 years, while in 2017 only 3.91% did not survive (Roser et al., 2019b). People live longer and healthier lives. They have become richer, work less and can afford more goods and services. The industrial revolutions have also had important economic, social and political changes. The level of urbanisation has grown – from less than 10% of the global population in 1800 to 55% in 2016 (Ritchie & Roser, 2019). The share of people working in agriculture, mining and fisheries decreased while the share of people in industries and services increased dramatically (Roser, 2013). The source of wealth changed – from owning land and resources, through owning factories, trains and computers, to owning data. The share of women in the workforce increased, and they were given voting rights. New political systems (e.g. imperialism, liberalism, social democracy) were established, and some of the dictatorial systems (communism, fascism, Nazism) crashed notoriously (Babb, 2018; Heywood, 2017).

In general, the preceding three industrial revolutions improved the economic well-being of populations and created a sufficient number of jobs in existing and new industries for workers displaced by automation. Although there had been some resistance to automation during the last two centuries by people who saw their livelihoods threatened by technology (Jones, 2006), the spread of automation was gradual; it took several decades for automation to unfold its impacts, thereby providing employees with sufficient time to requalify, to learn how to use the new technology and be competitive on the labour market. Societies had sufficient time to adjust to the new realities and develop economic, political and social responses to the challenges posed by automation. The current fourth Industrial Revolution based on robotics, artificial intelligence (AI), nanotechnologies, etc. (Schwab, 2016) is accelerating the automation of manufacturing and service delivery processes and it is qualitatively different from the preceding industrial revolutions. The previous three industrial revolutions did not eliminate humans from the economic processes. Companies still relied on labour because it was necessary to design and operate the machines and to make economic decisions. The greater involvement of women in the labour market at the beginning of the 20th century due to men's conscription into the armies of the First World War not only expanded the labour supply that companies could rely on but also increased households' disposable incomes (Ortiz-Ospina et al., 2018). The increase in the production output met an increase in consumption due to the growth of the population and the improved economic well-being of people. Although societies faced economic crises and high unemployment rates due to temporary misbalances between supply and demand such as the Great Depression of 1929–1933, economies succeeded in creating sufficient jobs to maintain low unemployment rates. This, however, may not be the case with the current and future industrial revolutions. In the future, we expect more processes to be completely automated and implemented without human intervention due to technological advances. Automated factories would not need human workers on production lines (Ross *et al.*, 2018; Solari *et al.*, 2022; Soroush *et al.*, 2020); chatbots and virtual/digital humans would take over customer service (Stoilova, 2021) and autonomous vehicles would eliminate drivers (Faisal *et al.*, 2019). Moreover, artificial autonomous agents (AAs) (Bösser, 2001) will take economic decisions thereby replacing humans not only as a production factor but also as decision-makers. In that way, the current economic system will be completely transformed into a robonomic system.

Robonomics is an economic system that relies on robots, automation technologies and AI to manufacture products, provide services and implement administrative processes in companies, organisations and public authorities (Ivanov, 2021). The role of humans in the robonomic economic system is minimised and limited to the overall design and control of the processes. The term 'robonomics' was coined by Crews (2016) and popularised by Ivanov (2017, 2021) in a series of publications. Various authors have used different terms to denote the automated economic system. For instance, Chase (2016) talks about 'economic singularity' when AI takes economic decisions instead of humans as a continuation of the 'technological singularity' concept of Vinge (1993). Brynjolfsson and McAfee (2014) refer to it as the 'second machine age', while Bootle (2019) names it the 'AI economy'. However, regardless of the term, the basic idea is the same – humans are replaced as a production factor by technology. In this book, we use the term 'robonomics' because of its simplicity and rising popularity.

As an economic system, robonomics has specific principles and characteristics, driving forces, benefits, challenges and solutions to them, as outlined in Table 1.1, which will be elaborated on in this chapter and Chapters 2 and 3. This chapter presents the key principles and characteristics of robonomics. It also delves into the macroenvironmental, microenvironmental, corporate level and psychological drivers of robonomics. Chapter 2 outlines the main short- and long-term benefits and challenges of robonomics as an economic system, while Chapter 3 critically evaluates some of the possible solutions to the challenges of robonomics. The subsequent chapters delve into various aspects of tourism and hospitality in an automated economy.

Table 1.1 Conceptual framework of robonomics as an economic system

Drivers	Advent	Characteristics
Macroenvironmental factors *Technology*: • Advances in Raia technologies *Demography*: • Ageing population *Politics*: • Governmental control on populations *Legal framework*: • Antidiscrimination laws • Labour laws • Taxation • Hygiene regulations • Safety and security laws *Biosecurity*: • Epidemics, pandemics *Culture and society*: • Attitudes towards Raia technologies **Microenvironmental factors** *Labour market*: • Lack of sufficient and qualified human employees *Competitive pressure*: • Adoption of Raia technologies by competitors *Customers*: • Acceptance of Raia technologies **Corporate level factors** *Economic efficiency*: • Cost-efficiency • Productivity • Operations management **Psychological factors** *Preferences*: • Managers' preferences towards the use of robots/automation technologies instead of human employees • People's desire not to work	• Adoption of Raia by individual companies in an industry • Spread of Raia among companies in an industry • Gradual spread of Raia among industries and countries • Spill over effects of Raia from developed to developing economies in the form of substituting low-cost labour in developing countries for automated factories in developed economies and bots	• Artificial autonomous agents are production factors and economic agents • Automated decision-making • Active use of a variety of single- and multi-purpose industrial, service and social robots • High level of automation of production • High cost-efficiency of production • High level of standardisation of services • Small and dispersed automated factories, close to consumers • Disconnection between employment and incomes • Labour and capital abundance are not sources of competitive advantages, but knowledge and creativity are • Fewer but more knowledge-intensive jobs

	Impacts		Solutions
	Benefits	Costs	
Short / mid term time frame	• Decrease in costs and prices • Improved environmental sustainability of production • Accelerated scientific research	• Unemployment and relative overpopulation • Lower salaries • Psychological problems of people who find themselves with too much free time, nothing to do and no need to work • Fear, social unrest and political instability • Migration • Wars	*Short and mid term (on the road to robonomics):* • Mandating employment • Government job creation • Work sharing • Employment impact statements • Tax policies • Financial incentives for job creation

Long term time frame	• Improved *quality of life* in the long term due to:
	  o People are liberated of hard and dangerous work
	  o Drastic increase of leisure time
	  o Improved quality of the environment
	  o Less stress
	  o Improved health
	  o Increased life expectancy
	• Global government and global citizenship
	• Accelerated space exploration
		• Possible functional illiteracy
		• Division of society between employed and unemployed
		• Changes in social values
		• Possible decrease in population in the long term
			Long term (during robonomics):
			• Constant and fluid free lifelong education
			• Entertainment
			• Tourism and leisure activities
			• Volunteering
			• Universal basic income
			• Robot-based taxation
			• Birth control / birth right patents
			• Redefinition of human rights

1.2 Principles and Characteristics of Robonomics

As an economic system, robonomics has several fundamental characteristics that distinguish it from the current and preceding economic systems in human history (Ivanov, 2017, 2021).

1.2.1 High level of automation

This is the most important characteristic of robonomics and all the other issues, benefits and challenges of robonomics are more or less connected to it. Automation replaces labour as a production factor not only for repetitive, dirty, dull and dangerous physical tasks but also for many cognitive and emotional tasks (Huang *et al.*, 2019). The majority of goods and services are produced or delivered by automation technologies rather than by human beings. Industrial robots manufacture goods (Ross *et al.*, 2018). Autonomous vehicles transport people and cargo (Dimitrakopoulos *et al.*, 2021). Virtual humans provide customer services (Burden & Savin-Baden, 2019). Intelligent automation helps organisations keep track of documentation (Bornet *et al.*, 2021), while digital assistants organise people's daily schedules (Bäuml *et al.*, 2020). Medical bots and wearable and implantable devices track people's health status, eating habits and physical activities and provide personalised recommendations (Bhunia *et al.*, 2015; Delabrida Silva *et al.*, 2018; Dey *et al.*, 2019; Velez & Miyandoab, 2019). Human microchip implants provide their holder's identification and access to their homes and public facilities (Ivanov *et al.*, 2014; Michael & Michael, 2013), while brain implants augment the cognitive capability of people (Cinel *et al.*, 2019; Raisamo *et al.*, 2019). Kiosks, holograms and robots serve tourists at hotels and restaurants (Belanche *et al.*, 2021; Ivanov & Webster, 2019a; Lee & Oh, 2022; Rastegar *et al.*, 2021; Seyitoğlu *et al.*, 2021). Humans interact with a variety of single- and multi-purpose service and social robots in their daily lives (Royakkers & Van Est, 2016) and even develop sexual relationships with robots (Danaher & McArthur, 2017; Lee, 2017; Zhou & Fischer, 2019). Automation is permeating every aspect of human life in a robonomic system. It is ubiquitous, pervasive and inescapable. Silicon has effectively replaced carbon as a production factor.

High-level automation of economic activities is associated with the high cost-efficient production. The technologies of the future would allow the economically efficient on-demand single/few unit(s) production of some goods. This would lead to a decrease in the efficient production capacity (EPC) of factories (see Figure 1.1). The EPC is when the increase in capacity does not lead to a significant drop in average production costs and their curve remains nearly horizontal. Thus, the company no longer enjoys economies of scale. Curve A in Figure 1.1 indicates the average costs per unit as a function of the production capacity of a company without automation. Its EPC is denoted by EPC_1 and it is between the

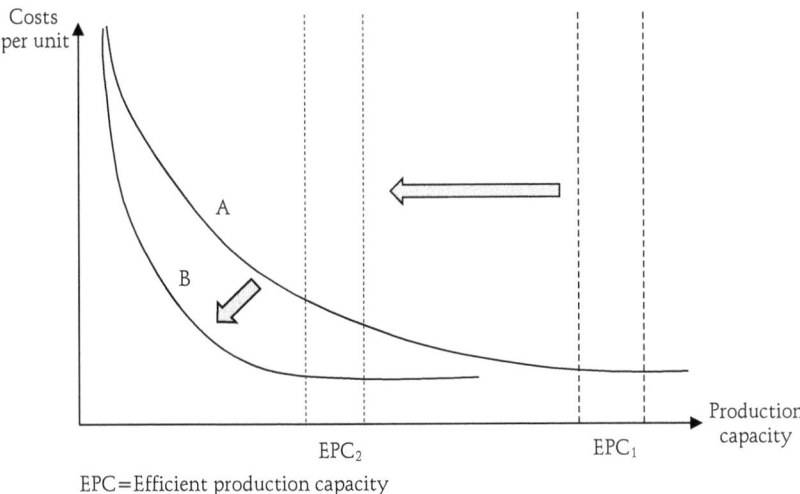

Figure 1.1 Changes in the efficient production capacity due to automation

two vertical dashed lines. The implementation of automation shifts the curve of the average costs from A to B while the EPC of a factory decreases from EPC_1 to EPC_2. As a consequence, factories will be smaller, more geographically dispersed and located closer to customers to decrease logistics costs. Mass production of standardised goods ('make-to-stock' manufacturing process) will be replaced by mass customisation and mass personalisation ('make-to-order') at similar or even lower costs (Da Silveira et al., 2001; Eyers & Dotchev, 2010).

Similar to manufacturing, the provision of services in robonomics is highly automated. Although technology in the future would advance sufficiently to allow the personalisation of services, humans would probably be more capable than technology in adapting the services to the desires and preferences of individual customers. Service companies in a robonomic system mainly provide cheap, technology-delivered, high-tech services or expensive, human-delivered, high-touch services; however, various shades of grey exist between these two extremes. Back-of-house operations are largely automated in all service companies, regardless of whether they offer high-tech or high-touch services, and rarely provide a significant competitive advantage or differentiation opportunity.

1.2.2 Artificial autonomous agents serve as economic agents

In robonomics, automation is not only a production factor; many economic decisions (e.g. purchases, use of resources, pricing, sales) are taken and implemented by artificial AAs. Bösser (2001: 1002) defines AAs as 'software programs which respond to states and events in their environment independent from direct instruction by the user or owner of the agent, but acting on behalf and in the interest of the owner'. AAs can exist as purely digital programmes such as chatbots and algorithms for trading on the financial markets, or they can be embedded in physical devices such as robots and autonomous vehicles. In robonomics, AAs serve as economic agents on the markets, similar to humans/households, companies and other organisations. They take economic decisions, often without direct human supervision but with significant economic repercussions for their owners. Thus, AAs' decisions to buy, sell or use resources, goods and services impact the economic well-being of their owners.

The relationship between AAs and humans in terms of decision-making includes six basic approaches presented in Table 1.2, namely: 'AA-out-of-the-loop', 'AA-on-the-loop', 'AA-in-the-loop', 'human-out-of-the-loop', 'human-on-the-loop' and 'human-in-the-loop'. The approaches are delineated based on an AA's and human's involvement in the decision-making process – recommending a decision to the other entity in the process, making the decision, overriding the decision of the other entity and the lack of opportunity to interfere in the decision-making process. The 'AA-out-of-the-loop' is currently the most widely used approach. In it, the decision is taken by a human although it may be implemented by an AA. The AA is not involved in any way in the decision-making process, e.g. preparing an offer for the organisation of an event at a hotel. In the 'human-in-the-loop' approach, the AA recommends a decision (e.g. changes to room rates) but the human has to decide whether or not to accept the recommendation (Ivanov, 2021). In the 'human-on-the-loop' approach, the AA takes and implements the decision but the human can override the decision. For instance, the revenue manager of a hotel chain can amend the prices set by the AA if she considers them to be too high or too low. In the 'human-out-of-the-loop'

Table 1.2 Decision-making processes

		Human			
		Recommends a decision to the AA	Takes the decision	Can override the decision of the AA	Cannot interfere in the decision of the AA
Artificial autonomous agent	Recommends a decision to the human		*Human-in-the-loop*		
	Takes the decision	*AA-in-the-loop*		*Human-on-the-loop*	*Human-out-of-the-loop*
	Can override the decision of the human		*AA-on-the-loop*		
	Cannot interfere in the decision of the human		*AA-out-of-the-loop*		

approach, the human is effectively eliminated from the decision-making process; decisions are taken entirely by the AA and they cannot be changed (e.g. high-frequency trading on the financial markets) (Ivanov, 2023). While the human 'in-', 'on-' and 'out-of-the-loop' approaches emphasise the human in the relationship as the implicitly superior entity, the AA 'in-' and 'on-the-loop' approaches emphasise the AAs as the superior decision-makers. In the 'AA-in-the-loop' approach, the human recommends a decision but it is up to the AA to accept it or not (e.g. a human generates various options for product bundles but the AA decides which ones to visualise to customers based on their expected likelihood to buy derived from customers' purchase history and search behaviour). In the 'AA-on-the-loop' approach, the human takes and implements the decision but the AA can override it (e.g. a personal digital assistant cancels the online purchase order of sweets for a customer who is overweight). Currently, these two approaches seem purely theoretical because they require humans to transfer their control over decision-making to AI. In the 'human-out-of-the-loop' approach, the transfer of control is complete but this can be justified for many routine and repetitive decisions with predictable outcomes (e.g. the decisions of a vacuum cleaning robot or an industrial robot) or when the decision requires speed that goes beyond the capabilities of the human brain to process information (high-frequency trade, military drones). However, there is a principle difference between the 'human-out-of-the-loop' approach and the AA 'in-' and 'on-the-loop' approaches. In the AA 'in-' and 'on-the-loop' approaches, the human is in a subordinate position compared to the AA as her recommendation ('AA-in-the-loop') or decision ('AA-on-the-loop') may not be considered or may be overridden by the AA. Hence, the authority of the human over the AA is challenged. In the robonomic system, most of the economic decisions will be in the human 'on-' and 'out-of-the-loop' approaches but the other approaches and a combination of approaches will also be applied. The greater involvement of AAs in economic decisions means that they would need to be considered as customers (Ivanov & Webster, 2017) or employees. This also raises the question of granting rights to AAs, especially to social robots (Gellers, 2021; Gunkel, 2018), as elaborated on in Chapter 3.

1.2.3 Few humans work

One of the consequences of the high level of automation, the cost-efficiency of production and the economies of scale provided by automation is that the robonomic system offers few jobs for human employees. The majority of people (probably over 80% of the working-age population) do not work, while a small share of humans are involved in highly paid jobs. The outsourcing of decision-making and production to technology decreases the need for human labour in companies and organisations and changes the nature of the work available to human employees (Webster & Ivanov, 2020a). They are involved in more strategic research, planning, monitoring, oversight, artistic and creative activities that require complex problem-solving, emotional intelligence and social skills rather than dealing with operational and repetitive tasks. The automation of manufacturing, administration and service delivery processes and the use of AAs that make independent economic decisions mean that economic value is not created by human labour but by automation. The Labour Theory of Value (Dooley, 2005) becomes irrelevant and is completely obliterated by automation. The low share of working humans creates various benefits and challenges that are elaborated on in Chapter 2.

1.2.4 Disconnection between employment and income

One of the main outcomes of the preceding characteristics of robonomics is the disconnection between employment and income due to the low level of employment. As Ivanov (2021) outlined, household incomes currently come from four different sources: investment/ownership of assets (i.e. dividends, interest payments, rent, capital gain); employment (i.e. salaries and wages, self-employment income); retirement accounts (i.e. pensions); and welfare (e.g. government payments to the unemployed, to people with special needs or other social groups). Employment-related income is, however, the most important income source that most households rely on. In robonomics, due to the low level of employment, the latter will not be the main source of income for households. Although a small share of people will work and will be highly paid, the overwhelming majority of people will not do so and thus will not receive employment income. As money is still the means of exchange in the robonomic system, people will need money to buy goods and services. The lack of employment means that societies would need to be creative and consider non-conventional strategies in income generation for the non-working populations, as discussed in Chapter 3.

1.2.5 Sources of competitive advantage

The high cost-efficiency of production, the high level of automation and the low EPC of companies mean that capital and labour abundance are not the sources of their competitive advantage but knowledge and creativity. This has already been recognised by many authors for the current economic system (Peters *et al.*, 2009; Urbancova, 2013; Wijaya & Suasih, 2020), but robonomics takes this to extremes. New start-ups would not need much cash to sustain operations or to build production facilities, something that brought many start-ups into the red. Rather, they have to focus on the innovativeness of their products, the value they create for the users and their products' synergies and compatibility with the existing digital ecosystem.

1.2.6 Überveillance

Überveillance or excessive surveillance (Michael & Michael, 2013) is an inherent characteristic of robonomics. Digital assistants, voice-activated devices, microchip implants, wearable tech and various other digital technologies constantly monitor and track all human activities – work related; purchase and consumption of goods and services; generated waste; communications and interactions with other people, robots, AAs and devices; eating, drinking, sex, hygiene and body health, etc. Everything is measured, recorded and creates a data point in personalised and big databases that feed AI algorithms for predictive analytics. Privacy is a term from the past. Überveillance is required to achieve the maximum efficiency of the robonomic economic and social systems, to achieve the fluent provision of personalised services (e.g. AI can track the communications of customers and visualise ads and offers to products that customers are likely to buy), to protect the health and safety of people (e.g. tracking infected people to minimise the spread of pathogens), to create a safe work environment for those people who still work, etc. Überveillance raises significant ethical issues related to privacy and behavioural control as in Chapter 3.

1.3 Drivers of Robonomics

The robonomic system may sound far-fetched and frightening to some readers, but many factors work towards the greater adoption of automation in society that would ultimately lead to robonomics (see Table 1.1).

1.3.1 Macroenvironmental factors

1.3.1.1 Technology

Advances in automation technologies are the most important driver of robonomics because they enable the automation of more and more tasks. Something previously impossible to automate is now automatable and will be completely automated in the future. The word 'robot' was invented only a century ago by Karel Čapek in his 1920 play R.U.R. (Rossum's Universal Robots) and robots were still part of science fiction 60 years ago. The first industrial robot was installed in 1961 (Stone, 2005) and, as of 2020, already 3,014,879 units have been installed worldwide with an average growth of 9% of new installations during the period 2015–2020 (IFR, 2021). In just six decades, robots have taken an important place in manufacturing. In the same year, over 20 million service robots helped households in their daily activities (Transforma Insights, 2021). And this phenomenal growth of robotics in recent years is largely due to the improved technological capabilities of robots. The same can be said for any other automation technology. The improved capabilities of automation technologies in the future will make it possible to automate tasks and replace humans as a production factor in activities that are currently perceived as human strongholds.

1.3.1.2 Demography

Previous studies have shown that demography is a major driver of robonomics (Webster, 2021) and the adoption of automation technologies in specific industries such as tourism and hospitality (Webster & Ivanov, 2020b). Since 1965, the average number of children per woman has been decreasing on a global scale and in many countries it is now below the replacement rate of 2.1 children per woman. It is 0.9 children in the Republic of Korea, 1.5 on average in EU countries and 1.7 in the United States and China (World Bank, 2021). Low birth rates lead to ageing and declining populations, and disruptions in the labour market (new entrants to the labour market cannot compensate for retiring workers), thus creating significant labour shortages. In practice, societies have three options to mitigate the negative impacts of demography on their economies: 'produce', 'import' or 'substitute' people (Ivanov & Webster, 2019a). Having more babies ('produce' strategy) is beneficial in the long term but requires several decades before newborns enter the labour market and the increased labour supply to unfold its impacts. Furthermore, many people may resist or not comply with a government policy of having more babies because they consider reproduction a personal choice. Relying on immigrants ('import' strategy) increases the labour supply in the short run but it creates social tensions when immigrants and their hosts belong to different cultures and religions. The solution that causes the least tension is to use automation technology ('substitute' strategy) to decrease the labour demand by automating tasks, processes and job positions for which companies find it difficult to employ human employees. It solves both short- and long-term labour shortages without causing a clash between groups of people belonging to different cultures and religious groups. According to forecasts from the United Nations (UN, 2019), the population of Asian, European and Latin American countries will decrease by 2100 compared to 2020 (in many countries by over 20%), while the population of Africa will significantly increase (it will more than double). Thus, the pressure on the labour markets of countries with declining populations is going to be more severe in the future, which will stimulate automation.

1.3.1.3 Politics

It is no secret that governments track citizens. CCTV and public surveillance cameras, face recognition technologies, microchips on identification documents (passports, driving licences and ID cards), mobile phones, email clients, web servers, digital assistants, voice-activated devices and other smart/digital technologies generate data that government agencies can use with the help of respective software to track terrorists and criminal activities. AI is actively used in surveillance (Feldstein, 2019). Often, government actions exceed the maximum necessary actions to fight terrorism and crimes and enter the area of massive control over populations (Feldstein, 2021). In 2021, for example, there were 117 public surveillance cameras for every 1000 citizens in Taiyuan, China; 90 in Wuxi, China; 73 in London; and 64 in Indore, India (Buchholz, 2021). Citizens are told that they have to sacrifice some of their liberties (e.g. privacy) for a safe and secure society (Moore, 2011). Massive control of populations is only possible with sufficiently developed automation technologies that provide cost-effective tracking of people. Thus, government policies towards greater control over populations inevitably lead to the wider adoption of automation and AI that facilitate the überveillance of people.

1.3.1.4 Legal framework

It may sound paradoxical but laws created to protect human employees at the workplace (e.g. labour laws, anti-discrimination laws, social security legislation) often have a negative impact on employment. For instance, companies may prefer to use automation rather than hire employees whom they would find difficult to fire, or to minimise potential discrimination lawsuits. Additionally, current tax legislation requires employers to make social security payments for their employees, which add costs on top of the salaries they pay; however, the use of robots and other automation technologies does not involve such taxes. Moreover, legislation can even require companies and organisations to use automation (in the institution theory by DiMaggio and Powell [1983] this is the 'normative' pressure). For example, hygiene regulations may require them to use self-service kiosks, touchless devices, cleaning and disinfection robots, etc. (Ivanov *et al.*, 2022). Furthermore, safety and security laws may force them to implement face recognition and tracking technologies to guarantee the safety of employees and customers and the security of their property.

1.3.1.5 Culture and social attitudes towards automation technologies

Previous studies have shown that there are significant cultural differences in the perceptions of and attitudes towards automation technologies (e.g. Bartneck *et al.*, 2005; Haring *et al.*, 2014) and that national culture shapes how people adopt new technologies (Lee *et al.*, 2013). Four decades ago, Sheridan *et al.* (1983) emphasised that some cultures (e.g. Japanese) are more receptive to automation because their social norms encourage productivity and quality, loyalty and devotion to the company/management, and focus on the long-term returns on investment rather than short-term profits. Companies and organisations in societies that are more open to new technologies, in general, and automation, in particular, would find it easier to automate their processes.

1.3.1.6 Biosecurity

Until 2020, biosecurity was overlooked as a driver of automation. The COVID-19 pandemic showed that biosecurity can serve as a short- and mid-term stimulator of automation (Ivanov *et al.*, 2022). The pandemic revealed that jobs in many service industries (hospitality, retail) were vulnerable and lost when the demand for these services disappeared due to the spread of the virus and the actions of governments to curb it, such as lockdowns and capacity limits. Many employees decided to move to other industries that provide greater job security. This created labour shortages in the affected industries that reacted in different ways – from increasing salaries through importing labour to automating processes. To mitigate the negative impacts of future pandemics on their business, automation will seem more attractive to managers than other options, especially in the context of ageing and declining populations.

1.3.2 Microenvironmental factors

1.3.2.1 Labour market

Labour shortages are one of the most important microenvironmental drivers of robonomics. They may be a result of various factors: demographic crisis (see Section 1.3.1.2), competition for labour by other industries, unattractiveness and low prestige of the industry (due to low salaries, long working hours, seasonality of employment, physical efforts required, difficult/dangerous/stressful working conditions, etc.), the vulnerability of jobs in the industry to external shocks such as viral outbreaks (see Section 1.3.1.6), and more. When companies face a lack of sufficient and qualified employees, they are forced to automate processes to decrease their dependence on labour and avoid the escalation of labour-related costs. Instead of competing for the limited labour supply, they re-engineer processes to decrease their labour demand, and automation is a key element of this re-engineering (Aversano *et al.*, 2002; Dey & Das, 2019). Thus, organisations have no feasible options to labour shortages other than to automate.

1.3.2.2 Competitive pressure

When competitors use automation and gain a competitive advantage due to cost savings and/or differentiation (Porter, 1995), companies feel the urge to follow their example. In the institution theory by DiMaggio and Powell (1983), competitors' actions form the 'mimetic' pressure. Following competitors' technology strategies allows companies to learn from the mistakes of their competitors and save costs by implementing technologies when their prices drop. However, such behaviour sacrifices the first-mover advantage and may not lead to a long-term sustainable advantage for the follower; it may even be counterproductive. As the Technology–Organisation–Environment framework (Tornatzky & Fleischer, 1990) emphasises, managers need to consider various organisation-level factors such as the size of the company and slack resources/production capacity before jumping into implementing new technologies rather than blindly copying competitors' actions. The positioning strategy of the company ('high-tech' or 'high-touch') also plays a role, especially in the service industries (Ivanov & Webster, 2019b; Seyitoğlu & Ivanov, 2020). For example, companies that rely on high-touch human contact in their service delivery may not need to automate front-of-house operations although they may automate most of the back-of-house tasks.

1.3.2.3 Automation technologies markets

Automation technologies are essentially products. Hence, their prices and product characteristics play a vital role in their adoption by companies, organisations and households. In recent years, the prices of automation technologies have steadily decreased. For instance, Ark Invest (2019) reports that the average price of an industrial robot more than halved during the period 2005–2017 – from US$68,659 in 2005 to US$27,074 in 2017. At the same time, labour costs increased. The decreasing prices of automation technologies, their improved characteristics and the rising labour costs due to labour shortages make automation technologies more affordable to companies and organisations and more price competitive compared to labour, stimulating their wider implementation.

1.3.2.4 Customers

When customers accept and expect to be served by a robot, chatbot, kiosk or other pieces of automation technology (this is the 'coercive' pressure in the institution theory by DiMaggio and Powell [1983]), it makes economic sense to introduce these in the service delivery process of a company/organisation provided these technologies fit its strategic positioning (see Section 1.3.2.2). On the contrary, when customers resist automation, automating front-of-house operations becomes challenging. As automation becomes more pervasive, people get used to it, and their expectations about services change; hence, in the future, there will be less resistance to the implementation of automation in the front-of-house operations of service companies although it will not disappear completely.

1.3.2.5 Partners

Partners can also exert coercive pressure and force companies/organisations to use automation. For instance, hotel chains can include room service robots or robots for pancakes at breakfast as part of their standard service offer. This will force individual member hotels (franchisees) to buy or rent robots. Similarly, distributors or suppliers may require the implementation of automation and full digitalisation in their supply chain; hence, they would require their partners to use specific digital/automation technologies. The latter would need to either comply and automate or not work with these distributors/suppliers. The greater the bargaining power of suppliers and distributors (Porter, 1980), the easier for them to impose the implementation of automation technologies on their partners.

1.3.3 Corporate-level factors

Economic efficiency in terms of cost-efficiency, productivity and the profitability of operations is a key driver of robonomics. When managers see that automation would improve the competitive positions of their companies/organisations on the market (by saving costs, generating revenue and/or creating more value for customers), and would increase the return on investment of the shareholders, they would be more likely to invest in automation. As Figure 1.1. shows, automation decreases the average costs per unit. It also increases the productivity of companies/organisations and decreases the time needed to serve one customer/manufacture one unit of a product compared to human-implemented processes. Automation improves the management of operations by eliminating the waste of time and unnecessary activities from processes. It makes automated processes more predictable and reliable compared to human-implemented processes. Therefore, the quest for economic efficiency would be successful when automation is part of a company's/organisation's arsenal.

1.3.4 Psychological factors

On the road to a (nearly) completely automated economic system, companies and organisations pass through a stage where they use collaborative automation technologies, i.e. human employees and automation technologies work together to deliver a service, manufacture a product or perform an administrative task. In economic terms, the use of automation technologies makes employees more productive because they can produce more products/serve more customers with the help of technology than before automation. This is the enhancement effect of automation (Ivanov & Webster, 2019b), but it will only happen if employees actually work with the respective automation technology. Paluch *et al.* (2021) find that employees have different willingness to collaborate with automation technologies. They may passively support or actively embrace collaborative service robots, but they may also passively resist or actively sabotage an organisation's efforts to implement service robots. Hence, when employees have a high willingness to collaborate with automation technologies, it will be easier for companies and organisations to automate various processes in their operations. Additionally, some managers may prefer to use robots/automation technologies instead of human employees. A recent survey of US managers by MindEdge and Skye Learning (2019) found that 65% of respondents would keep the current level of automation in their organisations even if the financial benefits do not materialise because they perceive that robots deliver higher quality work than humans. Hence, they would prefer to work with robots rather than human employees.

1.4 Conclusion

This chapter presented the key principles and characteristics of robonomics as an economic system and outlined some of its key macroenvironmental, microenvironmental, corporate-level and psychological drivers. The factors of the environment work in different directions but, on balance, they largely stimulate automation. In the future, the improved capabilities of automation technologies, their productivity, price and cost-efficiency compared to human employees, rising labour costs, lack of a sufficient and qualified labour supply, demographic crisis, etc., work towards the massive implementation of automation technologies in various industries until the current labour-based economic system is transformed into a robonomics system.

Robonomics will not happen overnight – it will be an emergent phenomenon. The advent of robonomics will be a gradual but probably not particularly slow process that will follow several stages:

- Stage 1: Adoption of automation technologies by individual companies in an industry.
- Stage 2: Spread of automation technologies among companies in an industry.
- Stage 3: Gradual spread of automation technologies among industries and countries.
- Stage 4: Spillover effects of automation technologies from developed to developing economies in the form of substitution of low-cost labour in developing countries for automated factories in developed economies and bots.

Different countries and even different industries are now at different stages on the path to robonomics and they are advancing at different speeds. In 2020, China, Japan, the United States, the Republic of Korea and Germany, for example, concentrated 76% of all industrial robot installations worldwide (IFR, 2021); thus, these countries have become the powerhouses of automated manufacturing. It is interesting to note that all five countries have birth rates that are below the replacement rate (2.1 children per woman); therefore, demographic trends already work in favour of automation in these countries. The growing population in developing countries maintains an abundant supply of labour. On the one hand, abundant labour keeps its price at a subsistence level; labour remains cheap compared to automation, which hinders the implementation of automation technologies. On the other hand, growing populations and abundant labour place the emphasis on governments' macroeconomic policies on creating jobs to decrease unemployment, minimise social tension and (often) avoid protests. Automation is not on the priority list of these governments. Nevertheless, the interconnectedness of the global economy means that developing countries will feel the impacts of automation in developed countries despite the low level of automation in their own economies.

As an economic system, robonomics has its benefits and challenges, which are the subject of Chapter 2.

References

Ark Invest (2019) Average cost of industrial robots in selected years from 2005 to 2017 with a forecast for 2025 (in U.S. dollars) [graph]. In *Statista*. See https://www.statista.com/statistics/1120530/average-cost-of-industrial-robots/ (accessed 4 January 2022).

Aversano, L., Canfora, G., De Lucia, A. and Gallucci, P. (2002) Business process reengineering and workflow automation: A technology transfer experience. *Journal of Systems and Software* 63 (1), 29–44.

Babb, J. (2018) *A World History of Political Thought*. Edward Elgar.

Bartneck, C., Nomura, T., Kanda, T., Suzuki, T. and Kennsuke, K. (2005) Cultural differences in attitudes towards robots. *Proceedings of the AISB Symposium on Robot Companions: Hard Problems and Open Challenges in Human–Robot Interaction*, 12–15 April 2005, University of Hertfordshire, Hatfield, UK (pp. 1–4). AISB. See https://ir.canterbury.ac.nz/bitstream/handle/10092/16849/bartneckAISB2005.pdf.

Bäuml, M., Jordan, H. and Miller, K.J. (2020) Broker protocol for automated event scheduling by virtual assistant. *Technical Disclosure Commons*. See https://www.tdcommons.org/dpubs_series/3765 (accessed 30 January 2022).

Belanche, D., Casaló, L.V. and Flavián, C. (2021) Frontline robots in tourism and hospitality: Service enhancement or cost reduction? *Electronic Markets* 31 (3), 477–492.

Bhunia, S., Majerus, S. and Sawan, M. (eds) (2015) *Implantable Biomedical Microsystems: Design Principles and Applications*. Elsevier.

Bootle, R. (2019) *The AI Economy: Work, Wealth and Welfare in the Robot Age*. Nicholas Brealey Publishing.

Bornet, P., Barkin, I. and Wirtz, J. (2021) *Intelligent Automation: Welcome to the World of Hyperautomation*. World Scientific Publishing Company.

Bösser, T. (2001) Autonomous agents. In N.J. Smelser and P.B. Baltes (eds) *International Encyclopedia of the Social & Behavioral Sciences* (pp. 1002-1006). Pergamon. https://doi.org/10.1016/B0-08-043076-7/00534-9

Brynjolfsson, E. and McAfee, A. (2014) *The Second Machine Age: Work, Progress, and Prosperity in a Time of Brilliant Technologies*. W.W. Norton & Company.

Buchholz, K. (August 23, 2021) The most surveilled cities in the world [digital image]. See https://www.statista.com/chart/19256/the-most-surveilled-cities-in-the-world/ (accessed 2 January 2022).

Burden, D. and Savin-Baden, M. (2019) *Virtual Humans: Today and Tomorrow*. CRC Press.
Chase, C. (2016) *The Economic Singularity: Artificial Intelligence and the Death of Capitalism*. Three Cs.
Cinel, C., Valeriani, D. and Poli, R. (2019) Neurotechnologies for human cognitive augmentation: Current state of the art and future prospects. *Frontiers in Human Neuroscience* 13, 13.
Crews, J. (2016) *Robonomics: Prepare Today for the Jobless Economy of Tomorrow*. CreateSpace Independent Publishing Platform.
Da Silveira, G., Borenstein, D. and Fogliatto, F.S. (2001) Mass customization: Literature review and research directions. *International Journal of Production Economics* 72 (1), 1–13.
Danaher, J. and McArthur, N. (eds) (2017) *Robot Sex: Social and Ethical Implications*. MIT Press.
Delabrida Silva, S.E., Rabelo Oliveira, R.A. and Loureiro, A.A.F. (eds) (2018) *Examining Developments and Applications of Wearable Devices in Modern Society*. IGI Global.
Dey, S. and Das, A. (2019) Robotic process automation: Assessment of the technology for transformation of business processes. *International Journal of Business Process Integration and Management* 9 (3), 220–230.
Dey, N., Ashour, A.S., Fong, S.J. and Bhatt, C. (eds) (2019) *Wearable and Implantable Medical Devices: Applications and Challenges*. Academic Press.
DiMaggio, J. and Powell, W. (1983) The iron cage revisited: Institutional isomorphism and collective rationality in organizational fields. *American Sociological Review* 48 (2), 147–160.
Dimitrakopoulos, G., Tsakanikas, A. and Panagiotopoulos, E. (2021) *Autonomous Vehicles: Technologies, Regulations, and Societal Impacts*. Elsevier.
Dooley, P.C. (2005) *The Labour Theory of Value*. Routledge.
Eyers, D. and Dotchev, K. (2010) Technology review for mass customisation using rapid manufacturing. *Assembly Automation* 30 (1), 39–46.
Faisal, A., Kamruzzaman, M., Yigitcanlar, T. and Currie, G. (2019) Understanding autonomous vehicles. *Journal of Transport and Land Use* 12 (1), 45–72.
Feldstein, S. (2019) The global expansion of AI surveillance. Working Paper. *Carnegie Endowment for International Peace*. See https://carnegieendowment.org/files/WP-Feldstein-AISurveillance_final1.pdf (accessed 27 December 2021).
Feldstein, S. (2021) Governments are using spyware on citizens. Can they be stopped? Carnegie Endowment for International Peace. See https://carnegieendowment.org/2021/07/21/governments-are-using-spyware-on-citizens.-can-they-be-stopped-pub-85019 (accessed 9 January 2022).
Frey, C.B. (2019) *The Technology Trap: Capital, Labor, and Power in the Age of Automation*. Princeton University Press.
Gellers, J.C. (2021) *Rights for Robots: Artificial Intelligence, Animal and Environmental Law*. Routledge.
Gunkel, D.J. (2018) *Robot Rights*. MIT Press.
Haring, K.S., Mougenot, C., Ono, F. and Watanabe, K. (2014) Cultural differences in perception and attitude towards robots. *International Journal of Affective Engineering* 13 (3), 149–157.
Heywood, A. (2017) *Political Ideologies: An Introduction* (6th edn). Palgrave.
Huang, M.H., Rust, R. and Maksimovic, V. (2019) The feeling economy: Managing in the next generation of artificial intelligence (AI). *California Management Review* 61 (4), 43–65.
International Federation of Robotics (IFR) (2021) Executive Summary World Robotics 2021 Industrial Robots. See https://ifr.org/img/worldrobotics/Executive_Summary_WR_Industrial_Robots_2021.pdf (accessed 28 August 2022).

Ivanov, S. (2017) Robonomics: Principles, benefits, challenges, solutions. *Yearbook of Varna University of Management* 10, 283–293.

Ivanov, S. (2021) Robonomics: The rise of the automated economy. *ROBONOMICS: The Journal of the Automated Economy* 1, 11.

Ivanov, S. (2023) Automated decision-making. *Foresight* 25 (1), 4–19. https://doi.org/10.1108/FS-09-2021-0183 (accessed 02 May 2017).

Ivanov, S. and Webster, C. (2017) The robot as a consumer: A research agenda. *Proceedings of the 'Marketing: Experience and Perspectives' Conference*, 29–30 June 2017, University of Economics-Varna, Bulgaria (pp. 71–79). See http://ssrn.com/abstract=2960824 (accessed 02 May 2017).

Ivanov, S. and Webster, C. (2019a) Conceptual framework of the use of robots, artificial intelligence and service automation in travel, tourism, and hospitality companies. In S. Ivanov and C. Webster (eds) *Robots, Artificial Intelligence and Service Automation in Travel, Tourism and Hospitality* (pp. 7–37). Emerald Publishing. https://doi.org/10.1108/978-1-78756-687-320191001

Ivanov, S. and Webster, C. (2019b) Economic fundamentals of the use of robots, artificial intelligence and service automation in travel, tourism and hospitality. In S. Ivanov and C. Webster (eds) *Robots, Artificial Intelligence and Service Automation in Travel, Tourism and Hospitality* (pp. 39–55). Emerald Publishing. https://doi.org/10.1108/978-1-78756-687-320191002

Ivanov, S., Webster, C. and Mladenovic, A. (2014) The microchipped tourist: Implications for European tourism. In A. Postma, J. Oskam and I. Yeoman (eds) *The Future of European Tourism* (pp. 86–106). Stenden University of Applied Sciences.

Ivanov, S., Webster, C., Stoilova, E. and Slobodskoy, D. (2022) Biosecurity, crisis management, automation technologies, and economic performance of travel, tourism and hospitality companies: A conceptual framework. *Tourism Economics* 28 (1), 3–26. https://doi.org/10.1177/1354816620946541

Jones, S.E. (2006) *Against Technology: From the Luddites to Neo-Luddism*. Routledge.

Lee, J. (2017) *Sex Robots: The Future of Desire*. Palgrave Macmillan.

Lee, E.M. and Oh, S. (2022) Self-service kiosks: An investigation into human need for interaction and self-efficacy. *International Journal of Mobile Communications* 20 (1), 33–52.

Lee, S.G., Trimi, S. and Kim, C. (2013) The impact of cultural differences on technology adoption. *Journal of World Business* 48 (1), 20–29.

Michael, M.G. and Michael, K. (eds) (2013) *Uberveillance and the Social Implications of Microchip Implants: Emerging Technologies*. IGI Global.

MindEdge and Skye Learning (2019) Robomageddon: Future of work study. See https://blog.skyelearning.com/future-skills (accessed 4 January 2022).

Moore, A.D. (2011) Privacy, security, and government surveillance: Wikileaks and the new accountability. *Public Affairs Quarterly* 25 (2), 141–156.

Ortiz-Ospina, E., Tzvetkova, S. and Roser, M. (2018) Women's employment. Our World in Data. See https://ourworldindata.org/female-labor-supply (accessed 9 January 2022).

Paluch, S., Tuzovic, S., Holz, H.F., Kies, A. and Jörling, M. (2021) 'My colleague is a robot': Exploring frontline employees' willingness to work with collaborative service robots. *Journal of Service Management* 33 (2), 363–388. https://doi.org/10.1108/JOSM-11-2020-0406

Peters, M.A., Marginson, S. and Murphy, P. (2009) *Creativity and the Global Knowledge Economy*. Peter Lang.

Porter, M.E. (1980) *Competitive Strategy: Techniques for Analyzing Industries and Competitors*. The Free Press. Reprinted in 1998 with a new introduction.

Porter, M.E. (1985) *Competitive Advantage: Creating and Sustaining Superior Performance*. The Free Press. Reprinted in 1998 with a new introduction.

Raisamo, R., Rakkolainen, I., Majaranta, P., Salminen, K., Rantala, J. and Farooq, A. (2019) Human augmentation: Past, present and future. *International Journal of Human-Computer Studies* 131, 131–143.

Rastegar, N., Flaherty, J., Liang, L. and Choi, H.C. (2021) The adoption of self-service kiosks in quick-service restaurants. *European Journal of Tourism Research* 27, 2709.

Ritchie, H. and Roser, M. (2019) Urbanization. Our World in Data. See https://ourworldindata.org/urbanization (accessed 9 January 2022).

Roser, M. (2013) Employment in agriculture. Our World in Data. See https://ourworldindata.org/employment-in-agriculture (accessed 9 January 2022).

Roser, M., Ortiz-Ospina, E. and Ritchie, H. (2019a) Life expectancy. Our World in Data. See https://ourworldindata.org/life-expectancy (accessed 9 January 2022).

Roser, M., Ritchie, H. and Dadonaite, B. (2019b) Child and infant mortality. Our World in Data. See https://ourworldindata.org/child-mortality (accessed 9 January 2022).

Ross, L.T., Fardo, S.W. and Walach, M.F. (2018) *Industrial Robotics Fundamentals: Theory and Applications*. Goodheart-Willcox Co.

Royakkers, L. and van Est, R. (2016) *Just Ordinary Robots: Automation from Love to War*. CRC Press.

Schwab, K. (2016) *The Fourth Industrial Revolution*. World Economic Forum.

Seyitoğlu, F. and Ivanov, S. (2020) A conceptual framework of the service delivery system design for hospitality firms in the (post-)viral world: The role of service robots. *International Journal of Hospitality Management* 91, 102661.

Seyitoğlu, F., Ivanov, S., Atsız, O. and Çifçi, I. (2021) Robots as restaurant employees: A double-barrelled detective story. *Technology in Society* 67, 101779. https://doi.org/10.1016/j.techsoc.2021.101779

Sheridan, T.B., Vámos, T. and Aida, S. (1983) Adapting automation to man, culture and society. *Automatica* 19 (6), 605–612.

Solari, L., Martinez, M., Braccini, A.M. and Lazazzara, A. (eds) (2022) *Do Machines Dream of Electric Workers?: Understanding the Impact of Digital Technologies on Organizations and Innovation*. Springer.

Soroush, M., Baldea, M. and Edgar, T.F. (eds) (2020) *Smart Manufacturing: Concepts and Methods*. Elsevier.

Stoilova, E. (2021) AI chatbots as a customer service and support tool. *ROBONOMICS: The Journal of the Automated Economy* 2, 21. See https://journal.robonomics.science/index.php/rj/article/view/21 (accessed 16 December 2021).

Stone, W.L. (2005) The history of robotics. In T.R. Kurfess (ed.) *Robotics and Automation Handbook* (pp. 1–12). CRC Press.

Tornatzky, L.G. and Fleischer, M. (1990) *The Process of Technological Innovation*. Lexington Books.

Transforma Insights (August 26, 2021) Number of personal assistance robots worldwide from 2020 to 2030 (in millions) [graph]. In *Statista*. See https://www.statista.com/statistics/1259870/personal-assistance-robots-worldwide/ (accessed 1 January 2022).

United Nations (UN) (2019) World Population Prospects 2019. United Nations, Population Division, Department of Economic and Social Affairs. See https://population.un.org/wpp/Download/Standard/Population/ (accessed 02 January 2022).

Urbancova, H. (2013) Competitive advantage achievement through innovation and knowledge. *Journal of Competitiveness* 5 (1), 82–96.

Velez, F.J. and Miyandoab, F.D. (eds) (2019) *Wearable Technologies and Wireless Body Sensor Networks for Healthcare*. The Institution of Engineering and Technology.

Vinge, V. (1993) The coming technological singularity: How to survive in the post-human era. *Vision-21. Interdisciplinary Science and Engineering in the Era of Cyberspace: Proceedings of a Symposium Sponsored by NASA Lewis Research Center and the Ohio Aerospace Institute*, Westlake, Ohio, March 30–31 (pp. 11–22). IICA Biblioteca Venezuela.

Webster, C. (2021) Demography as a driver of robonomics. *ROBONOMICS: The Journal of the Automated Economy* 1, 12.

Webster, C. and Ivanov, S. (2020a) Robotics, artificial intelligence, and the evolving nature of work. In B. George and J. Paul (eds) *Digital Transformation in Business and Society Theory and Cases* (pp. 127–143). Palgrave-MacMillan.

Webster, C. and Ivanov, S. (2020b) Demographic change as a driver for tourism automation. *Journal of Tourism Futures* 6 (3), 263–270. https://doi.org/10.1108/JTF-10-2019-0109

Wijaya, P.Y. and Suasih, N.N.R. (2020) The effect of knowledge management on competitive advantage and business performance: A study of silver craft SMEs. *Entrepreneurial Business and Economics Review* 8 (4), 105–121.

World Bank (2021) Fertility rate, total (births per woman). See https://data.worldbank.org/indicator/SP.DYN.TFRT.IN (accessed 1 January 2022).

Zhou, Y. and Fischer, M.H. (eds) (2019) *AI Love You: Developments in Human–Robot Intimate Relationships*. Springer.

2 Consequences of Robonomics

Stanislav Ivanov

2.1 Introduction

This chapter focuses on the consequences of robonomics. It explores the benefits and challenges of this automated economic system and discusses them in different time frames (short term and long term). The renowned American futurologist Roy Amara supposedly stated that 'we tend to overestimate the effect of a technology in the short run and underestimate the effect in the long run' (Ratcliffe, 2016) largely due to applying linear thinking to a non-linear phenomenon. In practice, the effects of automation technologies will need time and require a critical mass to unfold. Tens of thousands of service robots globally in hotels, restaurants, bank offices, shopping malls, airports, theme parks, schools and universities, and more, may seem like a huge number, but they will be lost in the ocean of tens of millions of human employees in these industries. The adoption of automation technologies by companies is driven by various factors, some elaborated on in Chapter 1. The diffusion of technologies in an industry, among industries and countries, is a gradual process (Rogers, 1962/1983). Existing technologies in companies and new technologies' compatibility with them, the costs of new and existing technologies, the relative performance of new and existing technologies and other factors need to be considered when making an investment decision (Ivanov & Webster, 2019).

The implementation of automation technologies in an economy requires time and money. Thus, the diffusion of automation technologies will happen more slowly than technological evangelists preach; they typically focus on the technological capabilities rather than the practicality of implementation. Therefore, their expected impact in the short run (a few years) will be much lower than expected. However, once a critical mass of automation technologies adopters has been reached, various intended and unintended direct and indirect effects of automation technologies will start to materialise. Often, these effects go beyond the direct impacts of automation on the production process, service delivery, prices, costs, efficiencies and more, and may transform the fabric of society by changing positively or negatively the well-being and economic welfare of large segments of the population and, as a consequence, cause political tension between the winners and the losers of automation. The long-run effects of automation technologies can be numerous, diverse, profound, difficult to foresee in advance and, therefore, easy to underestimate.

The following discussion of the short-run and long-run benefits and challenges of robonomics (Ivanov, 2017, 2021) needs to be viewed in the context of Amara's law mentioned above. Where the boundary between 'short run' and 'long run' lies is relative. Should it be a few years? A decade? A couple of decades or more? We do not have a solution. That is why, in this chapter, we do not make precise forecasts about *whether*, *when* and *how* the benefits and challenges of robonomics will appear. Instead, they are outlined and the potential effects are discussed. For practical reasons, the short-run effects include those benefits and challenges that would be among the first to happen because they are direct or close indirect effects of the massive implementation of automation technologies on a global scale. The long-term benefits and challenges may require a generation or more to materialise. Additionally, the delineation between the benefits and challenges is based on the effects of robonomics on society as a whole. While there are always winners and losers among stakeholders, some of the effects will have a *predominantly* positive impact on society (i.e. they are classified as 'benefits') while others will generally have a negative impact (i.e. 'challenges').

2.2 Benefits of Robonomics

2.2.1 Short-term benefits of robonomics

2.2.1.1 Decrease in costs and prices

One of the main benefits of robonomics as an economic system is the economic efficiency of production that can be achieved through the massive use of automation technologies (Ivanov, 2017). The elimination of unnecessary activities in production/service delivery processes, the 24/7 availability, the increased production/service capacity of companies, the decrease in labour costs and other factors contribute to the increased economic efficiency of production/service delivery which leads to lower costs per unit/served customer and, consequently, lower prices to consumers. The history of technology has already shown that the implementation of automation technologies improves production and the operations management of companies and significantly contributes to their productivity, efficiency and financial results (Frey, 2019). In the robonomic economic system, where artificial autonomous agents (AAs) and automation technologies will be widely used, these benefits will be even stronger. However, several factors will work against the decrease in costs and prices.

2.2.1.1.1 Polanyi's paradox

The term 'Polanyi's paradox' was introduced by David Autor (2014) based on Michael Polanyi's (1966: 4) observation that 'We know more than we can tell'. In every organisation there is codified and tacit knowledge. The former is the knowledge explicitly expressed in service operations and production manuals, standard operating procedures and other written documents in organisations that explain how processes need to be implemented, and elaborate on organisational structures, the flow of information, responsibilities and other issues. The tacit knowledge includes 'non-codified, disembodied knowhow that is acquired via the informal take-up of learned behaviour and procedures' (Howells, 1996: 92). This is the personal knowledge that employees gain through their work experience that is difficult to convey to others and to codify: e.g. sense of aesthetics, which product options to propose to a specific customer and which words or photos to use in a promotional brochure or in a social media post to maximise customer engagement and sales. The successful implementation of automation technologies requires codified knowledge that can serve as the basis for the development of process algorithms that could be automated. As Autor (2014: 135) eloquently summarised, 'engineers cannot program a computer to simulate a process that they (or the scientific community at large) do not explicitly understand'. Therefore, the more tacit knowledge in organisations, the fewer the opportunities for automation. Although the need to maintain high and consistent service/product quality, transparency of organisational processes, traceability of actions and decisions, reporting to government bodies, tax authorities and owners, franchise contracts, protection against legal claims and other factors force organisations to codify as much of their knowledge as possible, part of it always remains tacit and puts a limit on the automatability of tasks/processes. Consequently, costs may not decrease as much as managers initially expected before the implementation of automation.

2.2.1.1.2 Ironies of automation

Directly linked to the tacit knowledge from Polanyi's paradox are the 'ironies of automation'. According to Bainbridge (1983), automation may exacerbate rather than eliminate problems with the human operator of technology for two major reasons. The first one is designer errors, i.e. mistakes in the design of the automation technology that may decrease overall efficiency because it may actually require greater effort by the human employees. Restaurant employees might find a service robot too difficult to operate or its cleaning and maintenance might require too much of their time. While design errors are easily identified and corrected (although this may come at higher costs), the second reason is more important and may have a greater impact on the economic efficiency of automated processes. In a partially automated process, the human operator is left with an arbitrary collection of tasks because the engineers could not automate them (Bainbridge, 1983), or because these tasks are not repetitive enough or require *ad hoc* handling so they are not economically feasible for automation and need to be left to humans (see Van der Aalst *et al.*, 2018). This means that the tasks that human employees will implement may not be logically or sequentially connected. For example, some employees in a fast-food restaurant with self-ordering kiosks for customers may deal with cleaning the floors and the tables, and giving the ordered food and drinks to the customers, while the tasks related to the provision of information about the menu, ordering and payment are transferred to the kiosks. Therefore, the employees may not see how their work fits within the broader production/service process and their work motivation, productivity and efficiency may suffer. Additionally, humans may rely too much on automation (e.g. pilots relying on an airplane's autopilot system for landing), which will decrease their efficiency when they need to implement the task in case of automation failure. In the end, automation would increase the efficiency of the automated tasks but may decrease it for human-implemented tasks and the overall efficiency gain might not be what was initially expected.

2.2.1.1.3 Taxation of automation technologies

One of the solutions to the challenges of robonomics, which is discussed in Chapter 3, is the taxation of automation technologies. Regardless of how automation-related taxes will be calculated (e.g. flat tax per robot, kiosk, chatbot; VAT/sales tax surcharge; or in another way), their financial burden will be transferred from the companies to the customers through higher prices.

2.2.1.1.4 Market power of technology suppliers

The implementation of automation technologies provides a significant competitive advantage to companies. As the development of such technologies requires large financial resources and significant economies of scale, we could see a concentration of market share in a few suppliers of particular technologies and a long tail of small suppliers (many of which are start-ups) with small sales. For instance, in 2020, the largest supplier of industrial robots in terms of sales had double the sales of the second-largest supplier and 20 times the sales of the seventh-largest supplier (Statista, 2021). Although for some sectors (e.g. warehouse automation) such concentration was much lower (Interact Analysis, 2021), future mergers, acquisitions and defaults may increase the market power of a few large players. The rising start-up costs for new suppliers of automation technologies also serve as a barrier to potential competitors and favour the large incumbent firms. As a result of their dominant position, the prices of automation technologies may be significantly high, requiring organisations that use them in their operations to increase their prices or at least not to transfer the cost savings to their customers to recover their investments in automation technologies. Moreover, organisations that use automation technologies may find themselves in a vendor lock-in situation in which they face high switching costs and cannot easily change their supplier of automation technology (Ivanov & Webster, 2019). For instance, parts of the robots may not be entirely compatible or the transition to a new artificial intelligence (AI)-based pricing software may be accompanied by a loss of some of the data in the software that is currently used by the company. The vendor lock-in effect works towards the higher prices of automation technologies because it decreases the competition among their suppliers by decreasing the opportunities for customer migration. The technology-as-a-service (TaaS) model (renting rather than buying the technological product) allows the automation users to mitigate the vendor lock-in effect because TaaS decreases their technology investment and maintenance costs. Therefore, in the robonomic economic system, most organisations will not buy but rent technological products and software packages on a need-to-use basis. Access to technology will replace ownership of technological products.

2.2.1.2 Improved environmental sustainability of operations

The role of technology in the sustainable development of societies has been widely acknowledged (Behera & Prasad, 2020; Zacher, 2017). Specifically, automation technologies allow for the elimination of unnecessary activities/tasks in operations, the decreased use of resources, less waste and, therefore, improved environmental sustainability of operations. Movement detectors in corridors, for example, allow hotels to save on electricity because the corridors only light up when movement is detected, otherwise remaining dim. Smart boilers heat the water based on the consumption patterns of their users. Autonomous vehicles optimise their route and electricity consumption based on traffic data received from other connected vehicles (Vahidi & Sciarretta, 2018). AI helps companies make better sales forecasts (Sun *et al.*, 2008) and minimise waste due to overproduction.

When evaluating the impacts of automation technologies on the environmental dimension of sustainability, we need to consider the overall impact through their whole life cycle, i.e. manufacture, use, repair and disposal/recycling. All automation technologies use electricity and the implementation of new digital technologies (e.g. augmented, virtual and mixed reality) increases energy consumption. Therefore, a major element of automation's environmental impact is the sustainability of the electricity they use. In the future, greater adoption of sustainable electricity production (e.g. wind, solar, tidal, hydro, geothermal and nuclear power) rather than fossil fuel will contribute significantly to the environmental sustainability of automation technologies. Furthermore, the batteries of computers, tablets, smartphones, robots, appliances and electric vehicles are still expensive and toxic to produce and dispose of. Thus, the sustainability of automation technologies also demands the improved sustainability of batteries.

2.2.1.3 Accelerated scientific research

AI is already widely adopted in research in all fields of study from science, technology, engineering and mathematics (STEM) sciences to social sciences and the humanities (Cartwright, 2020; Cheng *et al.*, 2021; Egger, 2022). In the future, the massive implementation of AI will lead to accelerated scientific research that would benefit society in the long term through new research findings. However, this benefit of AI-based research does not require an automated economic system; hence, it will materialise regardless of the degree of automation in the economy and society. However, what is important here from the perspective of robonomics is the *automation of research* itself and the *automation of the publication of research results* (which one may consider the dream of academia). Artificial AAs will sift through vast amounts of data, perform the necessary analyses, make a scientific conclusion and produce reports in a standardised format to be used by human scientists. These reports can serve as the basis for academic publications of human scientists or they can be published as they stand. This raises questions about authorship rights because in practice the AAs' involvement in research would be sufficient to justify their (co-)authorship of the publications provided they were human beings. This issue has already been recognised by academia and publishers, and in April 2019, the first book written by an AI was published (Writer, 2019).

2.2.2 Long-term benefits of robonomics

2.2.2.1 Improved quality of life

The most important long-term benefit of robonomics is the improved quality of life. This benefit is not only an outcome of the automation of the economy, but it also needs to be the goal of political decision-makers. Quality of life will be improved in various ways, as elaborated on in Figure 2.1.

First, in the robonomic economic system, people are liberated from hard and hazardous manual work that requires significant physical efforts and is dangerous for human health. Tasks such as cleaning, lifting and moving heavy objects, work in the construction sector, agriculture, mining and chemical industries are already implemented with the help of technology, including robots, but future robots will take over even more

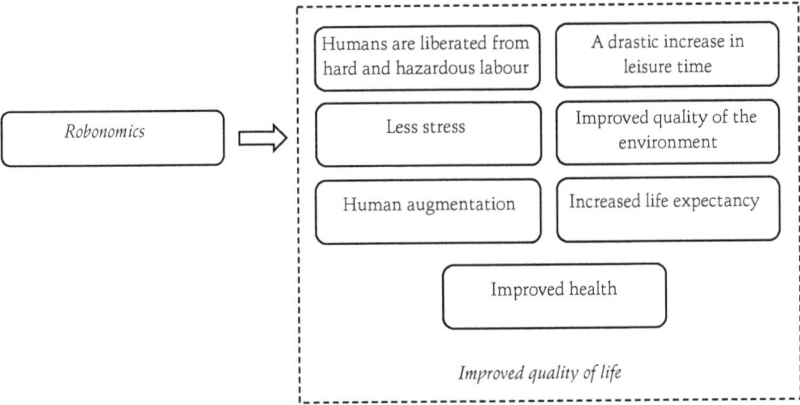

Figure 2.1 The positive impacts of robonomics on the quality of life

of these tasks. The pressure on the human body due to hard physical and dangerous labour will substantially decrease. Second, robonomics would lead to less work-related stress. There will be no more strict due dates, office politics, conflicts with managers, colleagues, customers or suppliers, no more sales targets to achieve and no more tedious paperwork. Work stress causes many health problems (Ganster & Rosen, 2013) that would be avoided in robonomics because automation could be successfully used to create a decent work environment (Tuomi *et al.*, 2020) for those humans who still work. Furthermore, highly intelligent digital assistants will take care of many tasks in the daily routine of people (organising trips, paying bills, renewing insurance, proposing options for presents for friends' birthdays, etc.) while robots and smart devices will take care of house chores and repairs, further decreasing the stress. Third, the massive implementation of automation technologies and the low share of working humans lead to a considerable increase in people's leisure time. Free time allows people to do more things without the need to choose between their preferences due to the lack of time. They can focus on more pleasurable, recreational, creative, intellectual and social activities they consider meaningful to them to achieve self-fulfilment and psychological harmony and connect with relatives and friends. Fourth, the increased sustainability of companies' operations means that automation technologies will improve the quality of the environment people live in. Less pollution in the robonomic economic system would help societies live in a healthier environment. Fifth, all these factors (the decrease in hard physical labour and work-related stress, the increase in leisure time and activities, etc.) inevitably lead to the better overall health of people and longer life expectancy. This would be further stimulated by human augmentation through a large variety of nanobots and wearable and implantable devices that monitor in real time the health status of people and take the necessary actions to improve it, e.g. nanobots that deliver drugs, remove cancer cells and blood clots (Li *et al.*, 2019, 2021; Suhail *et al.*, 2022). Augmented humans will be able to sense, process information and do things beyond the physiological limits of their natural bodies (Raisamo *et al.*, 2019). As a final result, people will have a longer and healthier life. A life expectancy above 100 years might be the new norm. Inevitably, there will be some negative impacts of automation on the quality of life, which are discussed further in this chapter.

2.2.2.2 Global government and global citizenship

One of the most important long-term benefits of the robonomic economic system is the unification of humanity, the creation of a world government and the establishment of global citizenship as opposed to our species' mosaic of national identities. The challenges that the massive implementation of robots, AI and other automation technologies raise, such as technological unemployment, political instability and migration (discussed in Section 2.3) require responses such as universal basic income (UBI) and the transformation of the tax system, which are feasible when implemented on a global scale (Lawhon & McCreary, 2020). Similar to environmental problems, the challenges of AI do not recognise national borders. Societies will be forced to undertake the next step in their evolution – unification. This process has been underway for many decades and the European Union is the best example of how countries that fought against each other for centuries and started two world wars can work together for the common good of their citizens. The European Union serves as the model for other regional blocs such as the African Union (Nubong, 2020). In the future, the challenges of automation would stimulate regional integration even further. In the long run, the regional blocs would be an interim step towards the unification of humanity into one global country run by a global government to solve global challenges. Naturally, there would be much resistance by nationalists who would prefer the partitioned political landscape; however, in time the support for the global unification of humanity would increase if it is shown to be effective towards the solution of global challenges related to the environment, human rights, the regulation of automation technologies, migration, security, etc.

2.2.2.3 Accelerated space exploration

Space is not only the last frontier of human exploration. The asteroids and the planets in the Solar system and beyond offer enormous natural resources for humanity's survival and development (Gregg, 2021; Pelton, 2016). They also offer places for the construction of new habitats that would transform humanity into a multi-planet species. But space is notoriously dangerous for humans. Weightlessness, radiation, extremely low temperatures and prolonged travel (due to the low speeds of spaceships and long distances to planets within our solar system and the nearest stars) make space travel dangerous for the health of astronauts and often practically inefficient at the current level of technological development. Although the next century will bring us an enormous technological process, *Star Trek* or *Star Wars* types of interstellar spacecraft that are faster than the speed of light are unlikely to be around. Passenger space travel will probably still be largely concentrated around the Earth, the Moon and Mars where permanent colonies could be established. Travelling to the gas giants (Jupiter, Saturn, Uranus, Neptune) and their moons requires many months of travel and the establishment of expensive space stations orbiting them to support humans. Currently, such endeavours are not financially feasible because of the enormous costs to construct a space station to support hundreds of people. Automation provides the solution. Using autonomous robots for much of the construction work and space exploration would save on costs because the space authorities would need to cater for fewer humans. Human astronauts require a lot of room in the space stations and the spaceships for sleep, work, hygiene, exercise, food supplies, etc. The systems of the space stations and spaceships need to create breathable air and clean the CO_2 and biological waste. The safety of humans is a priority. Using robots spares engineers these worries. Spaceships could carry more robots than humans due to better utilisation of room. Robots could be used in various environments that are hostile to humans. Robots are expendable, human astronauts are not. A robot failure would mean a loss of time, money and effort, the loss of an astronaut is a human tragedy. Due to the long distances between the Earth and the other planets in our solar system, the time to communicate with remote-controlled robots is often measured in hours. Thus, for the exploration of planets beyond Mars, robots need to be more sophisticated and make many if not all decisions autonomously without human control from Earth because the latter is impractical in the context of the time needed to communicate with the robot (Pyle, 2019). In short, space exploration by human astronauts and space tourism (Toivonen, 2021) are very romantic and inspiring, but the massive implementation of autonomous robots seems like the more (financially) feasible scenario for the coming decades.

2.3 Challenges of Robonomics

2.3.1 Short-term challenges of robonomics

2.3.1.1 Unemployment and relative overpopulation

The main negative consequence of automation pointed out by authors and politicians is technological unemployment. They fear that automation will eliminate jobs for humans who would remain unemployed, thus causing significant social tension. Thus, they look at the substitution effect of automation (replacing human employees) rather than its enhancement effect (helping employees be more productive) (Ivanov & Webster, 2019). Considering that employment is still a major source of income for most households worldwide, technological unemployment seems, in the eyes of many, as a justified reason for their negative attitudes towards automation technologies.

These automation fears have a long history (see Ivanov *et al.* [2020] for a detailed elaboration of previous studies on automation fears). As early as the 1810s, the Luddites in England looked at the new technologies in the textile industry as a threat to their jobs. In February 1928, a year and a half before the collapse on Wall Street and the start of the Great Depression, an article in *The New York Times* claimed that 'March of the machine makes idle hands. The prevalence of unemployment with greatly increased industrial output points to the influence of labour-saving devices as an underlying cause' (Clark, 1928). A couple of years later, Keynes (1930/1965) pointed out the possible technological unemployment of people but he looked at it as a temporary phenomenon. After the Second World War, the attitude towards automation was initially a little more optimistic. In the 1950s, when the first computers were already operational in the United States, Hurd (1958) discussed the new professions and job enhancement (which he called 'position enlargement') resulting from automation. In a similar vein, Rago (1965: 35) acknowledged the substitution effect of automation; however, according to him, this effect was not a reason for any fears among human employees because automation created more jobs through lower product prices and the creation of new jobs in areas 'where the "added" purchasing power generates "new" demand'. Later on, when automation technologies became more prevalent and sophisticated, and with the advent of robots and AI, the analysis of the social and economic impacts of automation on jobs divided authors into two groups. The first group (e.g. Brynjolfsson & McAfee, 2014; Danaher, 2019; LaGrandeur & Hughes, 2017) considers that the benefits of automation are larger than the challenges it causes and technological unemployment is not a big problem because the social systems will be adjusted to provide incomes not based on employment. The other group (e.g. Barrat, 2013; Frey, 2019; Leonhard, 2016) is more pessimistic about the new job creation and the viability of the social mechanisms devised to mitigate the negative consequences of technological unemployment.

As usual, more than 50 shades of grey exist between these two extreme positions. However, the improved technical capabilities of automation technologies mean that more and more tasks that are currently implemented by humans will become automatable. The decrease in the costs of automation technologies and the increase in their productivity relative to human employees mean that the automation of these tasks will not only be technologically possible but also economically efficient. Therefore, more and more tasks now performed by humans will be automated leading to the substitution of human employees for technology. Of course, automation will not only automate tasks but also create new tasks for humans. When the newly created tasks cannot compensate for the automated tasks, job positions are cut and people become unemployed. Hence, technological unemployment would depend on the balance between the substitution and enhancement effects of automation. Technological advances, operational efficiencies, costs and lack of sufficient and qualified employees in the labour market work towards greater automation (see Chapter 1). On the road to robonomics, technological unemployment will be inevitable because automation technologies will eliminate tasks for human employees faster than new tasks are being created. The problem, however, is not with technological unemployment *per se* but how quickly the social systems are being transformed to face the new economic realities. High and prolonged technological unemployment without the social net of an alternative income provision (e.g. in the form of a UBI) would stimulate a lot of tension in societies. It would give ground to neo-Malthusian politicians to declare that societies have reached the point of relative overpopulation, i.e. there are many more people than the economic system can employ and provide incomes for, and impose drastic population control. While relative overpopulation is a serious challenge opposite to the current demographic crisis faced by developed economies, it would be a real challenge if societies rely on employment as their main source of income and do not reorganise their economic systems to acknowledge the new technological realities.

2.3.1.2 Lower employment incomes

Previous studies have pinpointed that automation technologies have a negative impact on wages (DeCanio, 2016) because they decrease the demand for labour. While the demand for some job positions related to the planning, development, implementation and maintenance of automation technologies and the design of the respective operational processes will increase, the supply of such specialists on the labour market will increase as well. The impact on their salaries will be based on the balance between the supply and demand for their skills. The demand for other job positions may be much smaller than the potential supply, thus leading to lower salaries. Therefore, as a whole, the absolute amount and the share of employment income in households' budgets will decrease. At the same time, households may have revenues generated from artificial AAs they own – e.g. an algorithm trading on the stock exchange or an autonomous vehicle that works as a taxi. However, these incomes may not be sufficient to compensate for the loss of employment income, although they may be for some households.

2.3.1.3 Psychological problems of people

One of the main challenges on the road to and during robonomics is the abundance of free time. This may sound paradoxical because the lack of time seems to be a modern disease. As mentioned in Section 2.2.2.1, the drastic increase in people's leisure time will allow them to concentrate on other more pleasurable, recreational, creative, intellectual and social activities. While many people will successfully utilise their time, others may fall victim to what we call the 'Dutch disease of time abundance' – they will develop psychological problems because they have too much free time, no need to work and nothing to do. Previous studies have shown that the unemployed are more likely to experience psychological distress than the employed (Reneflot & Evensen, 2014). While currently unemployed people actively look for work and expect at some point to find work, in robonomics the majority of people would not look for jobs and would not expect to find a job because the labour demand would be low. This is the major difference – now the unemployment status of people is (with a few exceptions) temporary while in robonomics it might be for life. However, one factor mitigates this negative effect of free time. In the current economic system, people are expected to work to receive incomes while in robonomics, the majority of people will not be expected to work. Therefore, the feeling of guilt and frustration at being unemployed (Mikucka, 2014; Takahashi & Winefield, 2014) and the psychological pressure to find a job will be eliminated.

Another challenge with too much free time is that social contacts among people may suffer due to the lack of daily contact with colleagues at work. People may prefer to communicate with artificial AAs (robots, chatbots, digital assistants, virtual humans/holograms, etc.) rather than with other people because technology will provide seamless interactions with human users. Although most people will have a lot of leisure time to maintain their social relationships, the addiction to technology (Alter, 2017; Brooks *et al.*, 2020) may keep people hooked on playing games in virtual worlds, for example, rather than spending time speaking with fellow humans.

Therefore, the psychological distress of non-working people in a world without work might be greater than that of unemployed people in the current economic system. Social institutions would need to take care of the organisation of people's free time to avoid psychological problems – some specific examples will be elaborated on in Chapter 3.

2.3.1.4 Political instability

Technological unemployment and decreased employment incomes will inevitably lead to significant political tension due to the decreased economic welfare of people in the short run when social mechanisms to replace employment incomes have not yet been fully established. Social unrest, street protests, frequent changes of governments and the emergence of populist anti-tech neo-Luddist political parties that see technology as the source of humans' problems will accompany societies on the road to robonomics. Political instability, unemployment and lower incomes will turn many people against technological progress because they will see themselves as its victims. These challenges will be mitigated through various mechanisms such as UBI, which are discussed further in Chapter 3.

As outlined in Chapter 1, robonomics will be an emergent phenomenon. Its advent will be a gradual transformation of economies and societies rather than an overnight economic/social revolution. Some countries will be further ahead on the road to robonomics than others, but all countries will be affected. Countries that have introduced social safety nets (e.g. UBI) may experience an influx of migrants from countries that have not so done yet. Although UBI will initially apply only to citizens of a country, political pressure by some political parties and human rights non-governmental organisations (NGOs) will probably lead to the expansion of its scope to cover everyone currently residing in the country, which will make countries with such regulations very attractive to migrants. The current legal provisions and practices of the social security systems of developed economies towards refugees and migrants hint that these will likely be transposed in the UBI's regulations. However, high immigration may further nurture political instability as previous studies have already shown, especially when migrants come from culturally distant countries (Gebremedhin & Mavisakalyan, 2013). In extreme cases, political instability may lead to wars as ways to give vent to social pressure and divert attention from domestic problems.

2.3.2 Long-term challenges of robonomics

2.3.2.1 Functional illiteracy

A century ago, riding a horse was considered one of the basic skills of people, particularly men. Now, only a small share of people can ride horses. The need to learn this skill decreased with the wider adoption of automobiles. Now, horse riding is mostly a recreational activity (Dashper, 2019). In robonomics, the automation of the economy and many aspects of social life would eliminate the need for people to learn many of the skills we currently consider essential, and this would lead to possible functional illiteracy. Learning to drive a car would not be necessary due to the massive use of autonomous vehicles, most of which will not even have a steering wheel. Writing and reading may decrease in importance too due to voice-activated devices, speech-to-text and text-to-speech conversion software. Planning skills may weaken because personal digital assistants will take care of people's daily routines and meeting arrangements. Functional illiteracy would not only make people dependent on technology for the implementation of various tasks and activities, but it would also make them unemployable (Vágvölgyi *et al.*, 2016). When people rely too much on technology, they will start losing some of their skills and will be unemployable even if sufficient jobs are available. However, people will remain users of technology; hence, the design of technology needs to consider the declining skills of most users. Therefore, we will observe a self-nurturing vicious circle: automation technologies and artificial AAs will make people's lives easier, but their greater adoption would lead to the loss of skills, making people more dependent on these technologies. The loss of skills would stimulate the greater implementation of automation to compensate for the lack of sufficient skills in people, and this would further aggravate their dependence on technology. Although people will have a lot of free time to learn and upgrade their skills, many of them may not have the stimuli to do so because they would receive social payments (e.g. UBI, discussed in Chapter 3) that would be sufficient for a decent living standard.

2.3.2.2 Social divisions

Societies have always been, are and will be stratified. Genealogy, ethnicity, race, nationality, language, religion, gender, wealth, place of living, political party membership and countless other factors lead to legalised or tacit social divisions. Although the caste system, slavery and segregation are officially part of humanity's past, social divisions and antagonism between different groups of people still flourish. Rich vs poor, urban vs rural areas, liberals vs leftists vs right-wing parties, Christians vs Muslims vs other religions vs atheists, locals vs immigrants, East vs West, North vs South and other fault lines determine the edges of the social tectonic plates.

The robonomic system will increase social divisions. On the one hand, the robonomic system will mitigate many of the divisions. For instance, telemedicine will give residents of rural areas access to medical services that mostly residents of urban areas currently enjoy (Smith & Gray, 2009). Massive open online courses already provide children from disadvantaged communities with the opportunity to access lectures from leading universities which they may otherwise not have access to, although some studies say that they continue to be under-represented in online courses leading to the widening not closing of the educational gap between them and other children (Hansen & Reich, 2015).

On the other hand, robonomics will create new division lines in societies – e.g. employed vs unemployed and augmented vs unaugmented humans. As outlined in Chapter 1, in robonomics only a small share of people work. The majority of people, probably over 80%, do not need to work to sustain themselves. Those who work would receive very high salaries while those who do not will rely on the basic income they receive weekly/monthly. The differences in incomes and the way they contribute to the economy and society would lead to inevitable tension and antagonism between the two groups. Another important division would be between augmented and unaugmented humans. Augmented humans will have a variety of implantable devices, including brain microchip implants, which will allow them to be more productive if they work. They will be able to process much more information and do so much faster than unaugmented humans, they will be more physically fit and healthier and, ultimately, they will live longer. While some of the augmentations would be cheap, available and affordable to everyone, others will be expensive. The latter will lead to divisions even within the augmented humans group based on the their level of augmentation. Human augmentation will further exacerbate the division between employed and unemployed because most of the employed people will be augmented – augmentation will serve as a barrier to the transition between the two groups because it may be a prerequisite for employment. To some extent, there could be the development of a new form of racism, those who are augmented (since there will be visible manifestations of their augmentation) and an intersection with class (since the augmented will also tend to be wealthier).

2.3.2.3 Changes in human values

As with every social revolution, the advent of robonomics will inevitably lead to changes in human values. The value of time and time management will be challenged because people will have a lot of free time and AAs are taking care of the timeliness of all activities that involve humans. People will transfer the responsibility for managing their time to AAs. Additionally, many people may stop valuing work, innovativeness and creativity because they receive income without work. They may not have sufficient stimuli for self-development because technology is providing them with the goods and services they need, not their labour. Others may devalue face-to-face human interactions and prefer to interact with technology rather than with humans. The large-scale implementation of artificial wombs (Gelfand & Shook, 2006) may decrease the importance that people put on having children. This ultimately leads to questioning the value of human life and what it means to be human.

2.3.2.4 Possible long-term decrease in population

As elaborated on in Chapter 1, demography is one of the main drivers of robonomics (Webster, 2021), but the relationship between robonomics and demography is bidirectional. A direct consequence of the previous challenge of robonomics is the possible long-term decline in populations. Devaluing social contacts may harm sexual relationships and human reproduction in the long term. Some people may find it challenging to establish and maintain romantic relationships with humans and prefer relationships with AAs including sex robots (Danaher & McArthur, 2017), which would lead to fewer babies. Others may choose not to have children or have children later in their lives. Yet, a third group may have more children because of the social benefits of robonomics (discussed in Chapter 3), while artificial wombs and infertility treatment may allow people with reproductive problems to have children. However, their children may not be sufficient to compensate for the fewer children of other people. Hence, in the long run, the number of babies and consequently the population size may decrease, but this is not necessarily bad for humanity. Although from a biological perspective it would lead to a smaller genetic pool and less genetic diversity, it has some significant benefits when it happens in a robonomic system: social systems will need to cater for fewer people, the consumption of resources and the environmental impact of societies will be smaller. Artificial wombs (Gelfand & Shook, 2006) would allow societies to give birth to as many children as necessary and not to depend on individuals' decisions and preferences towards reproduction. Additionally, artificial wombs would allow better genetic testing and only the birth of children free of specific genetic diseases. In the long run, such an approach to human reproduction will improve the health of people and decrease the social costs of healthcare because previous studies have shown that a disproportionate share of healthcare costs goes on patients with genetic diseases (see Gonzaludo et al., 2019).

2.3.2.5 Überveillance, privacy and behavioural control

As outlined in Chapter 1, robonomics will be characterised by überveillance and loss of privacy. Automation technologies will be pervasive, tracking everything people do, eat, watch, listen to, wear, write, people they talk to, etc. As mentioned in Chapter 1, this would be necessary for the efficiency of the robonomic economic system but it raises significant ethical issues. Digital technologies will collect huge amounts of data that could easily be utilised for manipulation and behavioural control. While some directions of behavioural control (e.g. behaviour related to the sustainable use of resources, recycling, waste separation, etc.) seem justified in the name of the common good, the same technologies could be used for a technological dictatorship if proper democratic mechanisms are not established. A social credit system for behavioural control similar to the one experimented in China might be officially introduced (Liang et al., 2018).

The advent of robonomics will force societies to redefine privacy and personal space. Activities that we currently universally consider private (e.g. using a restroom and taking a shower) may no longer be considered private. Smart toilets may quickly analyse their content after usage while smart glasses in bathrooms may analyse people's skin to identify potential health problems. The use of such devices might be a prerequisite for the validity of medical insurance. Tracking people (e.g. with human microchip implants or other devices) will increase the opportunities to solve some criminal acts but it would also be used to limit the free movement of people (e.g. revoking access rights to various buildings and to autonomous vehicles). Self-censorship might be the usual behaviour on social networks. Although people may formally have the right to freely express themselves, knowing that everything is tracked will force many to impose self-censorship.

2.4 Conclusion

Robonomics will have numerous positive and negative impacts on society. Some of them will be experienced on the road to robonomics when economies are still adjusting to the new technological realities while others will be more significant after very high levels of the automation of economies have been reached. This chapter outlined some of the impacts; however, we acknowledge that there are numerous other impacts one can think of, some of which may have unexpected consequences. One thing, however, is sure – robonomics will lead to the profound transformations of societies and will be accompanied by many challenges. How to overcome these challenges is the topic of Chapter 3.

References

Alter, A. (2017) *Irresistible: The Rise of Addictive Technology and the Business of Keeping Us Hooked*. Penguin.

Autor, D.H. (2014) Polanyi's paradox and the shape of employment growth. In *Proceedings of the Re-evaluating Labor Market Dynamics Symposium*, Kansas City: Federal Reserve Bank of Kansas City, 21–23 August 2014 (pp. 129–177). See https://www.kansascityfed.org/Jackson%20Hole/documents/4538/2014Autor.pdf (accessed 25 April 2022).

Bainbridge, L. (1983) Ironies of automation. *Automatica* 19 (6), 775–779.

Barrat, J. (2013) *Our Final Invention: Artificial Intelligence and the End of the Human Era*. Macmillan.

Behera, B.K. and Prasad, R. (2020) *Environmental Technology and Sustainability: Physical, Chemical and Biological Technologies for Clean Environmental Management*. Elsevier.

Brooks, S., Wang, X. and Schneider, C. (2020) Technology addictions and technostress: An examination of the US and China. *Journal of Organizational and End User Computing*, 32 (2), 1–19.

Brynjolfsson, E. and McAfee, A. (2014) *The Second Machine Age: Work, Progress, and Prosperity in a Time of Brilliant Technologies*. WW Norton & Company.

Cartwright, H.M. (ed.) (2020) *Machine Learning in Chemistry: The Impact of Artificial Intelligence*. The Royal Society of Chemistry.

Cheng, Y., Wang, T. and Zhang, G. (eds) (2021) *Artificial Intelligence for Materials Science*. Springer.

Clark, E. (1928, February 26) March of the machine makes idle hands. *The New York Times*.

Danaher, J. (2019) *Automation and Utopia: Human Flourishing in a World without Work*. Harvard University Press.

Danaher, J. and McArthur, N. (eds) (2017) *Robot Sex: Social and Ethical Implications*. MIT Press.

Dashper, K. (2020) Holidays with my horse: Human–horse relationships and multispecies tourism experiences. *Tourism Management Perspectives* 34, 100678.

DeCanio, S.J. (2016) Robots and humans: Complements or substitutes? *Journal of Macroeconomics* 49, 280–291. https://doi.org/10.1016/j.jmacro.2016.08.003

Egger, R. (ed.) (2022) *Applied Data Science in Tourism: Interdisciplinary Approaches, Methodologies, and Applications*. Springer.

Frey, C.B. (2019) *The Technology Trap: Capital Labor, and Power in the Age of Automation*. Princeton University Press

Ganster, D.C. and Rosen, C.C. (2013) Work stress and employee health: A multidisciplinary review. *Journal of Management* 39 (5), 1085–1122.

Gebremedhin, T.A. and Mavisakalyan, A. (2013) Immigration and political instability. *Kyklos* 66 (3), 317–341.

Gelfand, S. and Shook, J.R. (eds) (2006) *Ectogenesis: Artificial Womb Technology and the Future of Human Reproduction*. Rodopi.

Gonzaludo, N., Belmont, J.W., Gainullin, V.G. and Taft, R.J. (2019) Estimating the burden and economic impact of pediatric genetic disease. *Genetics in Medicine* 21 (8), 1781–1789.

Gregg, J. (2021) *The Cosmos Economy: The Industrialization of Space*. Springer.

Hansen, J.D. and Reich, J. (2015) Democratizing education? Examining access and usage patterns in massive open online courses. *Science* 350 (6265), 1245–1248.

Howells, J. (1996) Tacit knowledge. *Technology Analysis & Strategic Management* 8 (2), 91–106.

Hurd, C.C. (1958) The social problem of automation. In *Proceedings of the Western Joint Computer Conference: Contrasts in Computers*, IRE-AIEE-ACM 1958, Los Angeles, 6—8 May 1958 (pp. 13–16). https://doi.org/10.1145/1457769.1457774

Interact Analysis (2021) Market share of the leading companies in warehouse automation in 2020 [graph]. In *Statista*. See https://www.statista.com/statistics/1259245/warehouse-automation-companies-market-share/ (accessed 25 April 2022).

Ivanov, S. (2017) Robonomics: Principles, benefits, challenges, solutions. *Yearbook of Varna University of Management* 10, 283–293.

Ivanov, S. (2021) Robonomics: The rise of the automated economy. *ROBONOMICS: The Journal of the Automated Economy* 1, 11.

Ivanov, S. and Webster, C. (2019) Economic fundamentals of the use of robots, artificial intelligence and service automation in travel, tourism and hospitality. In S. Ivanov and C. Webster (eds) *Robots, Artificial Intelligence and Service Automation in Travel, Tourism and Hospitality* (pp. 39–55). Emerald Publishing. https://doi.org/10.1108/978-1-78756-687-320191002

Ivanov, S., Kuyumdzhiev, M. and Webster, C. (2020) Automation fears: Drivers and solutions. *Technology in Society* 63, 101431. https://doi.org/10.1016/j.techsoc.2020.101431

Keynes, J. (1930/1963) Economic possibilities for our grandchildren. In J.M. Keynes (ed.) *Essays in Persuasion* (pp. 358–373). WW Norton & Co.

LaGrandeur, K. and Hughes, J.J. (eds) (2017) *Surviving the Machine Age. Intelligent Technology and the Transformation of Human Work*. Palgrave Macmillan.

Lawhon, M. and McCreary, T. (2020) Beyond jobs vs environment: On the potential of universal basic income to reconfigure environmental politics. *Antipode* 52 (2), 452–474.

Leonhard, G. (2016) *Technology vs. Humanity*. Fast Future Publishing.

Liang, F., Das, V., Kostyuk, N. and Hussain, M.M. (2018) Constructing a data-driven society: China's social credit system as a state surveillance infrastructure. *Policy & Internet* 10 (4), 415–453.

Li, P., Lee, G.H., Kim, S.Y., Kwon, S.Y., Kim, H.R. and Park, S. (2021) From diagnosis to treatment: Recent advances in patient-friendly biosensors and implantable devices. *ACS Nano* 15 (2), 1960–2004.

Li, S., Jiang, Q., Ding, B. and Nie, G. (2019) Anticancer activities of tumor-killing nanorobots. *Trends in Biotechnology* 37 (6), 573–577.

Mikucka, M. (2014) Does individualistic culture lower the well-being of the unemployed? Evidence from Europe. *Journal of Happiness Studies* 15 (3), 673–691. https://doi.org/10.1007/s10902-013-9445-8

Nubong, G. (2020) The relevance of the European Union integration experience to the African Union's integration process. In S.O. Oloruntoba and T. Falola (eds) *The Palgrave Handbook of African Political Economy* (pp. 1069–1086). Palgrave Macmillan.

Pelton, J.N. (2016) *The New Gold Rush: The Riches of Space Beckon!* Springer.

Polanyi, M. (1966) *The Tacit Dimension*. Doubleday.

Pyle, R. (2019) *Interplanetary Robots: True Stories of Space Exploration*. Prometheus Books.

Rago, L.J. (1965) Tooling up for automation hampered by myths and fears. *Business & Society* 5 (2), 33–36. https://doi.org/10.1177/000765036500500206

Raisamo, R., Rakkolainen, I., Majaranta, P., Salminen, K., Rantala, J. and Farooq, A. (2019) Human augmentation: Past, present and future. *International Journal of Human-Computer Studies* 131, 131–143.

Ratcliffe, S. (2016) Roy Amara. In S. Ratcliffe (ed.) *Oxford Essential Quotations*. Oxford University Press. See https://www.oxfordreference.com/view/10.1093/acref/9780191826719.001.0001/q-oro-ed4-00018679 (accessed 22 April 2022).

Reneflot, A. and Evensen, M. (2014) Unemployment and psychological distress among young adults in the ORDIC countries: A review of the literature. *International Journal of Social Welfare* 23 (1), 3–15.

Rogers, E.M. (1962/1983) *Diffusion of Innovations* (3rd edn). The Free Press.

Smith, A.C. and Gray, L.C. (2009) Telemedicine across the ages. *Medical Journal of Australia* 190 (1), 15–19. https://doi.org/10.5694/j.1326-5377.2009.tb02255.x

Statista (2021) Leading companies in the global industrial robot market in 2020, based on revenue (in million euros) [graph]. In *Statista*. See https://www.statista.com/statistics/257177/global-industrial-robot-market-share-by-company/ (accessed 25 April 2022).

Suhail, M., Khan, A., Rahim, M.A., Naeem, M., Fahad, M., Badshah, S.F., Jabar, A. and Janakiraman, A.K. (2022) Micro and nanorobot-based drug delivery: An overview. *Journal of Drug Targeting* 30 (4), 349–358.

Sun, Z.L., Choi, T.M., Au, K.F. and Yu, Y. (2008) Sales forecasting using extreme learning machine with applications in fashion retailing. *Decision Support Systems* 46 (1), 411–419.

Takahashi, M. and Winefield, A.H. (2014) Mental health of the unemployed in Japan. In M.F. Dollard, A. Shimazu, R.B. Nordin, P. Brough and M.R. Tuckey (eds) *Psychosocial Factors at Work in the Asia Pacific* (pp. 231–251). Springer Science + Business Media. https://doi.org/10.1007/978-94-017-8975-2_12

Toivonen, A. (2021) *Sustainable Space Tourism: An Introduction*. Channel View Publications.

Tuomi, A., Tussyadiah, I., Ling, E.C., Miller, G. and Lee, G. (2020) x=(tourism_work) y=(sdg8) while y= true: automate (x). *Annals of Tourism Research* 84, 102978.

Vahidi, A. and Sciarretta, A. (2018) Energy saving potentials of connected and automated vehicles. *Transportation Research Part C: Emerging Technologies* 95, 822–843.

van der Aalst, W.M., Bichler, M. and Heinzl, A. (2018) Robotic process automation. *Business & Information Systems Engineering* 60 (4), 269–272.

Vágvölgyi, R., Coldea, A., Dresler, T., Schrader, J. and Nuerk, H.C. (2016) A review about functional illiteracy: Definition, cognitive, linguistic, and numerical aspects. *Frontiers in Psychology* 7, 1617. https://doi.org/10.3389/fpsyg.2016.01617

Webster, C. (2021) Demography as a driver of robonomics. *ROBONOMICS: The Journal of the Automated Economy* 1, 12.

Writer, B. (2019) *Lithium-Ion Batteries. A Machine-Generated Summary of Current Research*. Springer. https://link.springer.com/content/pdf/10.1007%2F978-3-030-16800-1.pdf (accessed 11 April 2019).

Zacher, L.W. (ed.) (2017) *Technology, Society and Sustainability*. Springer.

3 Solutions to the Challenges of Robonomics

Stanislav Ivanov

3.1 Introduction

The development of technologies is faster than the development of societies and the human psyche. Günther Anders names this asynchronisation between humans and products/technology as the Promethean gap (Fuchs, 2021). It includes 'gaps between the relations of production and ideology, production and imagination, doing and feeling, knowledge and conscience, the machine and the body, production and needs' (Fuchs, 2021: 153). On the road to robonomics and during the period of the robonomic economic system, this gap may grow into a Promethean abyss due to the challenges of automated economies and societies (elaborated on in Chapter 2) and slow institutional changes due to societal resistance that will lead to the growing inadequacy and irrelevance of these institutions.

Governments and international organisations have already started to acknowledge the benefits and challenges of the robonomic economic system. In 2016, the Executive Office of the President of the USA, National Science and Technology Council and the Committee on Technology published three documents dedicated to the preparation of the American economy for the massive implementation of artificial intelligence (AI) and robotics, namely: Artificial Intelligence, Automation, and the Economy (2016), Preparing for The Future of Artificial Intelligence (2016) and The National Artificial Intelligence Research and Development Strategic Plan (2016). The Organisation for Economic Co-operation and Development published an extensive report on the impacts of automation on governments and companies (OECD, 2017); the World Economic Forum released reports on the future of jobs (WEF, 2023); and the European Union is expected to introduce an Artificial Intelligence Act. While all these efforts are commendable and needed, the challenges of robonomics will require coordinated action by different stakeholders (international organisations, governments, local authorities, producers of automation technologies, companies that implement automation in their operations, educational institutions, non-governmental organisations [NGOs] and individuals) on various levels (global, international, national, regional, local, corporate, household) (Ivanov, 2021). These actions need to be taken before the advent of robonomics to soften its social price.

This chapter elaborates on some of the instruments to eliminate or mitigate the challenges of robonomics. It looks at two groups of instruments – short- and mid-term instruments whose effectiveness will be highest during the road to robonomics when societies are transitioning from human- to automation-based economies, and long-term instruments that would be most applicable when economies are largely automated. It should be emphasised that societies would advance many other solutions beyond those elaborated on in the chapter, but the discussed instruments would form the backbone of societal mechanisms to solve or at least mitigate the challenges of robonomics.

3.2 Short- and Mid-Term Instrument (On the Road to Robonomics)

This group includes instruments that governments may use to temporarily curb the negative impacts of technological unemployment and provide the affected stakeholders with time to adjust. Stevens and Marchant (2017) name them 'incremental solutions' because they embody the incrementalism in public policy. According to Jones and Baumgartner (2005: 325), incrementalism 'implies that policy choice at a particular time is a marginal adjustment from a previous policy choice'. Therefore, the instruments in this group do not aim to provide a long-term solution to the challenges of robonomics but are used to solve or mitigate them on an *ad hoc* basis without significant changes to current public policy. These instruments include (Ivanov, 2017):

- Mandating employment: Government regulations that directly mandate companies to hire people for specific job positions (e.g. companies with above a certain number of employees to have a diversity manager or a sustainability manager) or indirectly create demand for certain job positions (e.g. an accreditation coordinator in a university). This solution protects the *status quo* and hinders companies from adjusting to changes in the environment triggered by technological change. Moreover, some of the required job positions may be completely irrelevant and unnecessary for companies because they do not add value to the competitiveness, profitability and work climate of organisations but are added only because they are legally required. Thus, this instrument creates jobs at the cost of decreased economic efficiency.
- Work sharing: Two employees share the same job position at reduced working hours. For example, instead of one person working for eight hours, two people work for four hours each. The advantage of this instrument is that more people will be employed and receive some employment income rather than social benefits. It is also convenient

for students, mothers and other people who prefer to work part time rather than full time due to other commitments. At the same time, work sharing does not provide employees with sufficient income for a decent quality of life and they may need to take other (part-time) jobs, which will not increase the actual number of employed people. Additionally, their engagement with the job may be low because they will perceive it as temporary, which may impact on their job performance due to the positive link between employee engagement and job performance (Ismail et al., 2019).
- Employment impact statements: These elaborate on the likely effect on jobs before the implementation of automation technologies. They can be prepared by companies that produce automation technologies and by companies that implement automation in their operations. In the first case, the employment impact statement will focus on the effects of this technology in general, based on the expected number of sold units, while in the second case the analysis needs to assess its specific effects on a particular company. The direct effects are relatively easy to determine and include the substitution of some job positions for automation, the creation of new job positions related to automation (e.g. programming, maintenance) and the transformation of a job position (changing the nature of work and, respectively, the skills requirements towards the employees). The indirect effects are very difficult to calculate because they include jobs created in companies supplying and maintaining automation technologies; job losses in consumer goods companies due to lower purchases by displaced workers from companies that implemented automation technologies; new jobs in other companies and industries where the workers displaced by automation move to and compensate for labour shortages, etc. The net direct quantitative effects of automation on jobs in a company is most likely negative but it may be (partially) compensated by a net positive indirect effect. Nevertheless, in the long term, the massive implementation of automation technologies will have a negative impact on jobs.
- Government job creation: This is a Keynesian and socialist approach to full employment. When technologies displace workers, governments step in as employers of last resort and hire them if no other available options exist to keep them employed. Governments may launch various infrastructural, environmental or urban revitalisation projects that will employ many workers (Stevens & Marchant, 2017), similar to the approach in the United States, Germany and other countries in the 1930s to battle the impacts of the Great Depression. On the one hand, this instrument is very politically attractive because it allows governments to quickly and substantially decrease unemployment, improving the living environment of local residents and the image of politicians, thereby attracting voters. On the other

hand, this instrument raises the question of its long-term funding, especially in the case of increasing technological unemployment that will decrease tax revenue to finance these projects.
- Financial subsidies for job creation: These are an alternative to government job creation. The government does not hire the displaced workers but instead pays companies to keep the workers by covering the social security and income tax payments of the employees and probably part of their salary. By doing so, from an accounting point of view, the government increases the cash inflows of companies and decreases their employee-related costs. This instrument was widely used during the COVID-19 pandemic (Costa Dias et al., 2020).
- Tax policies: They have the same goal as the financial subsidies for job creation but instead of paying companies, the government gives them tax reductions or tax credits to keep already hired employees or for new hires. Hence, this decreases the cash outflow of companies and also decreases the tax revenue of governments.

The common denominator of these short- and mid-term instruments to the challenges of robonomics is the nature and logic of government intervention. Jobs are created or, at least, not eliminated through legal regulations and fiscal stimuli, not because purely economic logic requires it. Therefore, the newly created or saved jobs are not necessarily economically efficient and may not contribute economically to society. For example, the fiscal stimuli provided by government may not be compensated for by the tax and social security revenue from these same jobs but by taxes on economically efficient companies and jobs. Therefore, although subsidised jobs create short-term employment for people, they are not sustainable in the long term and create fiscal problems for governments. Nevertheless, politicians now and in the future will be eager to subsidise jobs to attract voters due to their short-term focus on instruments that give quick results to attract votes during elections regardless of the long-term consequences of implementing these instruments. Additionally, the outlined solutions stimulate bureaucracy both in public authorities to control businesses and in companies to adhere to government regulations. This bureaucratic burden increases the costs to companies which need to be transferred to consumers through higher prices, and the government expenses which need to be covered by higher taxes, loans (that need to be repaid) or increasing the money supply (leading to inflation). In all cases, the companies, employees and consumers suffer.

3.3 Long-Term Instruments (Under Robonomics)

3.3.1 Constant, fluid and free lifelong education

Education is the natural solution to technological unemployment (Peters et al., 2019). The pace of technological changes quickly outdates the knowledge and skills gained at schools, colleges and universities. People will need to regularly upgrade their knowledge and skills to keep them relevant to the new technological realities. AI will be actively used to evaluate people's knowledge and skills, identify gaps in them, design personalised courses, generate the content of courses, deliver the content via digital avatars, evaluate the improvements in the knowledge and skills, etc. (Chen et al., 2020; Chiu et al., 2022; Xie & Wang, 2023). It will be constant and fluid education and robonomics will make lifelong learning a reality. The traditional divisions of higher education degrees into Bachelor's, Master's and Doctorate, together with Foundation and Associate degrees, Postgraduate Certificates and Postgraduate diplomas in some countries, may disappear or be shortened in duration due to the quick depreciation of the knowledge and skills for which graduates are certified. Instead, a stack of temporary skills certificates may replace them. People will need to renew them every few years, forcing them to participate in courses to refresh and upgrade their skills. At the same time, higher education institutions and training organisations will offer a multitude of other courses that are not related to the employability of people but to their well-being and quality of life – courses on languages, general knowledge (history, geography, politics, culture, biology, physics, etc.), hobbies, lifestyle, cooking, creative writing, music (playing musical instruments, composing, history of music), dances, etc. In robonomics, the cost efficiencies of AI will allow most AI-developed and delivered courses to be offered for free, which will eliminate the financial barriers that people face to improve their knowledge and skills. Because of the changes, education will be revolutionised and become something almost unrecognisable from what is now standard practice.

3.3.2 Entertainment, tourism and leisure activities

Contrary to people's desires, free time will be their biggest enemy. As discussed in Chapter 2, robonomics may cause a 'Dutch disease of time abundance' and psychological problems for people with too much free time that they do not utilise. To maintain the social fabric and avoid unrest, governments and organisations will need to benefit from the well-tested Roman formula *Panem et circenses* ('Bread and entertainment'). Entertainment, tourism and leisure activities will keep people's brains occupied – travelling to various destinations, concerts, movies, games, hiking, fishing, sports, events, practicing hobbies and countless other activities will help people fill the gaps in their time. Light drugs might need to be legalised as they already are in some countries (Bean, 2010) to allow people to have additional sensory experiences. Technologies will offer unlimited opportunities for such activities and experiences (Flavián et al., 2019; Neuhofer et al., 2014; Velasco & Obrist, 2020). Immersive online video games in the metaverse, implanting holiday memories (similar to the 1990 movie *Total Recall*); watching movies through the eyes of different characters; connecting to robotic avatars to explore destinations (including space and underwater destinations) instead of travelling to them (as in the 1982 movie *Blade Runner*) (Khatib et al., 2016); dream menus at homes, hotels, planes or cruise ships that will allow people to choose the theme of their dreams when they sleep; digital companions that will provide human-level interaction (Chandra Kruse et al., 2023); sex robots (Fan & Cherry, 2021; Peeters & Haselager, 2021), and many other technological solutions will create appealing and immersive experiences that fill people's time. The cost efficiency and the capabilities of technologies mean that these experiences will be created quickly, cheaply and mostly with the help of generative AI. They will be personalised to the preferences of individuals who will have an unlimited supply of technology-(co)created experiences at hand.

3.3.3 Volunteering

Volunteering will play the same role as entertainment, tourism and leisure but in a different way. While entertainment, tourism and leisure in robonomics will focus on hedonism, volunteering similar to education will focus on developing a sense of purpose, meaning and goal in life (Thoits, 2012). People will be involved in various volunteering organisations (formal or informal) and will dedicate time to solving or mitigating the consequences of health, environmental, social, ethical, economic and other problems faced by societies. Volunteering will not be a CV-building exercise but a pathway towards self-development and the search for a personal mission, growth and service beyond self by contributing to society without the expectation of a monetary reward. It will provide a path to a meaningful, healthy and sensible life. Volunteers will also help some of the side effects of the solutions to the challenges of robonomics such as potential health problems (obesity, deteriorated vision, lack of social contacts) due to excessive immersive digital experiences, ethical issues related to the redefinition of human rights or granting robots and other autonomous agents legal rights, among others. However, considering that many people might not want to give back to society, volunteering might be mandated by governments as a form of service requirements for citizen although they will choose what and when to volunteer.

3.3.4 Universal basic income

Universal basic income (UBI) is one of the most widely discussed solutions to the challenges of robonomics (Caputo, 2012; McDonough & Bustillos Morales, 2020; Pereira, 2017, 2023; Sheahen, 2012). In essence, under UBI every citizen of a country receives a monthly payment/stipend from the government regardless of their employment status, income and wealth. The amount might be different for adults and children. Similar concepts have been discussed by economists since the end of the 18th century, such as 'territorial dividend', 'social dividend' or 'negative income tax' (Bidadanure, 2019). UBI serves as a safety net for people because it will be 'universal' meaning nearly all will receive some income even if they fail in their entrepreneurial activities or lose their jobs. Therefore, UBI may stimulate many people to start new ventures because the negative consequences of potential failure are lower (they will still have financial resources to sustain themselves and their families). UBI is also easy to administer and may replace the complex social welfare systems that countries currently have and their extensive and expensive bureaucracy (Straubhaar, 2017). At the same time, UBI may discourage some people from work and they will exit the labour market. In a situation when companies are massively implementing automation technologies, this would not be such a bad impact because it will decrease the pressure on the labour market by decreasing the labour supply. Moreover, UBI may discourage crime because it will eliminate the lack of financial resources as a motive for crime. For instance, Deshpande and Mueller-Smith (2022) show that children removed from supplemental security income benefits are more likely to be involved in criminal activities, i.e. there is an inverse relationship between welfare and crime. In the context of robonomics, the study by Deshpande and Mueller-Smith (2022) implies that the provision of UBI may decrease some criminal activities thereby improving the social environment and quality of life of people.

UBI is not without its disadvantages. It is expensive to finance and needs to be combined with appropriate tax reforms (see Section 3.3.5 on robot taxes) because taxes on labour will not be a major source of revenue for governments. UBI may be financed in various ways (for details see Pereira, 2017, 2023). First, it would lead to budget cost saving due to replacing complex welfare systems and the associated costs to manage them. Second, the decreased budget revenue from employment (income taxes, social security taxes) may be partially compensated for by taxes on financial capital, land, raw materials (e.g. crude oil, natural gas and metals) and corporate profits; however, it is not yet clear how reliable these funding sources will be and they may have some unexpected side effects on markets and economies. For example, taxes on land or raw materials might be irrelevant in a highly automated circular economy. Moreover, most conceptual and empirical studies on the topic are based on the current economic system that relies on human labour and current tax systems. In a robonomic economic system when the overwhelming majority of people will not work and much of the budget revenue sources (taxes on employment) are not available or very limited, governments need to find other ways to fund the UBI scheme. Of course, printing money to cover UBI expenses will result in inflation. A potential solution is to introduce an expiry date on the digital funds that people receive under UBI, similar to the miles in airline loyalty programmes. For example, UBI receivers can spend the money within three months or another period after receiving it without the option to accumulate it for longer periods. This will limit the amount of UBI-related savings and the amount of money in the economy. When a person is borne, his/her parents start receiving UBI while the person receives UBI for life after adolescence. Money enters into circulation when a person is borne and leaves circulation when a person dies. The amount of money in circulation will be pegged to the number of people, i.e. the gold standard is replaced by a demographic standard. Of course, future research needs to delve deeper into the operational details and effects (positive and negative) of such a demography-based monetary policy and its feasibility.

Another major issue of UBI is that if it is not introduced on a global scale (e.g. by the United Nations) and administered by a global government, it will stimulate migration from countries without UBI to countries with UBI. Countries with well-developed welfare systems have already witnessed immigration from other countries (so called 'the welfare magnet hypothesis') (Agersnap et al., 2020). Therefore, the implementation of UBI needs to be either on a global scale (best option) or complemented by strict migration policies. Although migrants will not be entitled to UBI, many may move to countries with UBI in the hopes that their children borne in the host country will receive citizenship and be entitled to UBI. Migration flows will challenge the free movement of people and create tension with locals, as already observed (Burgoon & Rooduijn, 2021). Despite its disadvantages, UBI remains one of the most widely discussed instruments to offset the negative impacts of automation.

3.3.5 Robot-based taxation

Taxes on robots and other automation technologies have begun to gain significant academic attention (Guerreiro et al., 2022; Thuemmel, 2023). Such taxes serve two purposes: (a) make automation technologies more expensive for companies compared to labour and (b) compensate for the lost labour-based tax revenue in government budgets (income taxes, social/health security payments) due to the labour substitution effect of automation. Robots and other automation technologies are politically attractive taxable assets because people, especially in the case of high technological unemployment, will perceive them as a threat to their jobs and taxes on robots will be acceptable. Additionally, the owners of robots are few (therefore, few voters) while the people competing with robots for jobs are many (thus, many voters). Moreover, being physical long-term assets for their owners, robots might be taxed in a similar way to other assets such as real estate, especially since robots are essentially a form of capital that actively and directly participate in creating value for the market.

Although lucrative, the concept of robot taxes suffers from significant drawbacks. Much of the automation is actually achieved via other technologies (e.g. software bots, virtual/augmented reality and generative AI) rather than physical/tangible robots. Hence, the concept of robot-based taxation needs to be expanded to include all automation technologies. In practical terms, the tax legislation needs to provide very specific legal definitions of robots, automation technologies and other assets that will be taxed (Kovacev, 2020). However, this means that robotics/automation manufacturers may need to make only some small changes to their products to fall beyond the legal definition of taxable automation assets, so designers will be under pressure to design such technologies with loopholes in mind. After doing so, the legislators will need to amend the legal definitions to incorporate the new product versions and the cycle begins again. Therefore, the practical challenges of implementing robot/automation taxes may make them unfeasible.

A possible solution is the use of a combination of total assets tax and a transaction tax. The total assets tax is calculated as a percentage (e.g. 1% or less) of the total assets of a company/person/household. The more assets a company has, the larger the amount of tax it pays. Considering that from an accounting and financial point of view automation technologies are assets, the implementation of automation in a company will increase its tax base and, respectively, its assets tax. At the same time, if the company is not using automation but human employees, its assets on the balance sheet will not increase and its total assets tax base will remain unchanged. A small assets tax will also stimulate taxpayers (companies and individuals) to find productive ways to invest their money rather than lose money due to the assets tax. While the total assets tax may work for robots, production lines and other large equipment, it will not be effective for software bots, e.g. software for trading on the financial markets. Software is an intangible asset and is included on the balance sheet if it is to be used for a minimum number of years and is above a minimum value depending on countries' tax regulations. However, the software-as-a-service (SaaS) concept means that the software will not belong to the company and will not be included on its balance sheet; hence, it cannot be taxed as an asset of the company that uses it. Instead, a transaction tax may work better for software automation. For example, a certain percentage of each transaction (e.g. 1% or less), regardless of the nature of the transaction (sale/purchase, software licence fees, payment of salary, currency exchange, transfers between a company's own bank accounts, etc.) goes as a tax. Therefore, this transaction tax is similar to the proposed Tobin tax on short-term currency exchanges (Tobin, 1974/2015) and a financial transactions tax (Grahl & Lysandrou, 2014) but it is comprehensive because it includes all transactions, not only those related to currency exchange or trading financial assets. If a company is using software for algorithmic trading on financial markets

(Dimov, 2022), it will generate many transactions that will serve as the tax base for the transaction tax. The total assets tax and the transaction tax are easy to calculate, require little bureaucracy to administer and do not favour some economic subjects at the expense of others. When used in combination they can generate tax revenue from taxing automation technologies as assets (total assets tax) or the transactions and transfers associated with them (transaction tax).

3.3.6 Birth control/birth right tax

Eliminating the burdens and stress of work and providing the financial security of UBI might encourage many families to have numerous children, potentially causing a population surge. This could lead to societal challenges such as increased numbers of people without jobs, more individuals receiving UBI and escalating financial costs to maintain the living standards of people. Consequently, a neo-Malthusian birth control may be introduced like China's one-child rule that was in force between 1980 and 2016. For instance, families might be required to obtain a birth permit before conceiving. Without this, the pregnancy might be mandated for termination. The permit itself might be free of charge or paid with an exponentially increasing tax for the next child (i.e. birth right tax). This could deter many families from having more than one or two children but raises many ethical issues. For instance, in the case of the birth right tax, people with more financial resources will be able to legally have more children because they will be able to pay for the birth right. However, considering the plummeting global birth rates and the potential further decrease in the population in robonomics (as discussed in Chapter 2), this tool in the arsenal of policymakers may not need to be used although it will remain available.

3.3.7 Redefinition of human rights

The redefinition of human rights is one of the unconventional solutions to the challenges of robonomics. It is based on the premise that human rights are negotiated in society, they are flexible and change depending on the circumstances. Currently, the constitutions of all countries stipulate the rights to have children, to vote and to participate in economic activities as undisputable human rights. However, technological advances on the road to robonomics will challenge this notion and will show that this is not necessarily the case. A person may or may not have a particular right. In robonomics these two options for each right create eight unique combinations between the biological right to reproduce, the political right to vote and the economic right to receive UBI, presented in Table 3.1 with different social impacts (Ivanov, 2017).

Taken alone, each combination leads to mass poverty, country default or demographic crisis – all are equally dire. For example, in

Table 3.1 Redefinition of human rights in robonomics

Situation	Biological right to reproduce	Political right to vote	Economic right to receive UBI	Social impact
1	Yes	Yes	Yes	Country default
2	Yes	Yes	No	Mass poverty
3	Yes	No	Yes	Country default/social unrest
4	Yes	No	No	Mass poverty/social unrest
5	No	Yes	Yes	Demographic crisis
6	No	Yes	No	Demographic crisis
7	No	No	Yes	Demographic crisis
8	No	No	No	Demographic crisis

Source: Based on Ivanov (2017).

[1] people can have as many children as they wish and receive UBI for themselves and their children. Hence, people will have stimuli to vote for populist politicians who promise an increase in UBI and greater social benefits, as we already see in modern politics (see also Straubhaar, 2017: 79). In the long term, this leads to a country's default because the budget expenses may not be compensated by revenue. Situation [3] is similar to [1] but the lack of voting power would lead to social protests similar to those in the 18th, 19th and 20th centuries if UBI is insufficient to sustain a good quality of life. Situations [2] and [4] would lead to mass poverty because there would be no UBI as a social safety net against technological unemployment. Situations [5]–[8] lead to a demographic crisis due to strict birth control. However, a stable solution is formed by a combination between [2] and [7]: some people self-select (e.g. at a certain age) to receive UBI without the right to vote and have children (situation [7]) while others have the right to vote and reproduce but do not receive UBI, relying instead on their own economic activity and employment. This solution would definitely trigger an accusation of fascism and raise many yet to be answered ethical questions: e.g. Why should these three rights be considered together rather than independently? Why these three and no other rights? Why not everyone has all three rights? How can we be sure that people understand the consequences of their decisions when they choose one of the two options? What are the long-term impacts of this policy on the diversity of a population's genetic pool and the health of future generations? A softer version of this solution is when the choice between [2] and [7] is made not at a certain age but after the first child of the person. In that way, the genes of the person would remain in the genetic pool of the population. Additionally, the biological right to reproduce and the political right to vote in elections might not be rights that people receive at birth but rights that people earn through service to society (e.g. through volunteering).

3.3.8 Robot/AI rights

One of the topics that has received significant attention in recent years and that will form some of the legal fundamentals of robonomics is the rights of robots, in particular, and artificial autonomous agents, in general (Gunkel, 2018a, 2018b; Schwitzgebel, 2023; Tigard, 2023). The discussion revolves around whether robots and artificial autonomous agents can and should have rights, and if yes – which rights exactly. Figure 3.1 presents the robot rights matrix and the four situations in it based on Gunkel (2018a). Research so far has not reached a consensus about which quadrant of the matrix legislation needs to follow, each option attracting supporters.

Some of the arguments include (Gunkel, 2018a, 2018b):

- Robots (and artificial autonomous agents) cannot and should not have rights because they are just tools.
- Robots (and artificial autonomous agents) can and should have rights because they have some intelligence and autonomy and these will increase in the future.
- Robots (and artificial autonomous agents) can but should not have rights in order to protect humans and social institutions.
- Robots (and artificial autonomous agents) cannot but should have rights to avoid being mistreated by humans.

Despite the variety of philosophical arguments for and against robot rights, from a pragmatic point of view, in a robonomic economic system where artificial autonomous agents take economic decisions, sell to and

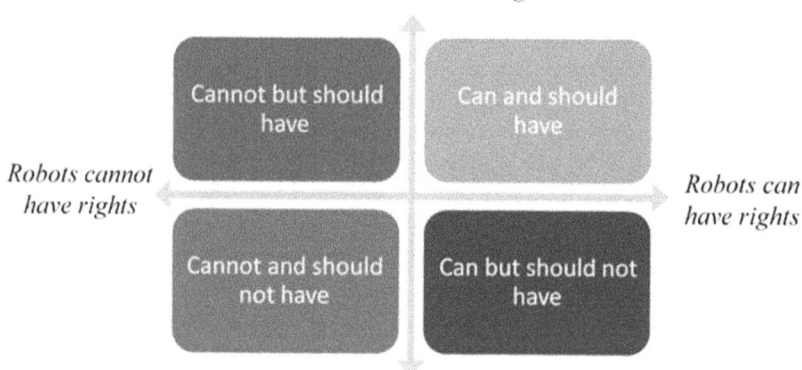

Figure 3.1 Robot rights matrix (Source: Developed by the author based on Gunkel, 2018a)

buy from other artificial autonomous agents (Ivanov, 2022), they need to be provided some limited economic rights, e.g. to own assets or to sign contracts, which will pave the way for autonomous companies entirely run (and some owned!) by artificial autonomous agents. At the same time, artificial autonomous agents may not have other rights, such as the right to vote, to avoid potential political complications. Currently, non-human legal entities such as organisations have economic rights independent of the humans that own or work in them while national and international legislation regulates various aspects of animal welfare. Therefore, granting rights to robots and artificial autonomous agents is not unthinkable and may evolve from existing practices.

3.4 Conclusion

This chapter outlined some of the instruments that societies can use to solve or mitigate the negative impacts of robonomics. As the chapter has shown, none of the instruments is perfect – all have their strengths and weaknesses, but when used in combination the strengths of some instruments may offset the weaknesses or further enhance the strengths of other instruments (e.g. combining UBI with total assets tax and transactions tax). Policymakers need to perform a comprehensive cost–benefit analysis of the implementation of each instrument and assess their direct and indirect short-term and long-term effects on societies, economies, demography, environment, etc.

The list of instruments is not comprehensive and will change in time with the evolution of technologies, legal systems and ethics. Societies will need to establish new institutions, rules, norms, relationships, mechanisms and reward systems that do not currently exist, and reshape or phase out existing ones. Many of the instruments may seem far-fetched, fascist or unrealistic, but many of them seem mandatory to curb the challenges of robonomics, guide human behaviour and avoid social unrest. To paraphrase Carl Sagan's famous aphorism that 'extraordinary claims require extraordinary evidence', in robonomics 'extraordinary challenges require extraordinary solutions'.

References

Agersnap, O., Jensen, A. and Kleven, H. (2020) The welfare magnet hypothesis: Evidence from an immigrant welfare scheme in Denmark. *American Economic Review: Insights* 2 (4), 527–542.

Artificial Intelligence, Automation, and the Economy (2016) Published by the Executive Office of the President of the USA. See https://obamawhitehouse.archives.gov/sites/whitehouse.gov/files/documents/Artificial-Intelligence-Automation-Economy.PDF (accessed 14 October 2016).

Bean, P. (2010) *Legalising Drugs: Debates and Dilemmas*. Policy Press.

Bidadanure, J.U. (2019) The political theory of universal basic income. *Annual Review of Political Science* 22, 481–501.

Burgoon, B. and Rooduijn, M. (2021) 'Immigrationization' of welfare politics? Anti-immigration and welfare attitudes in context. *West European Politics* 44 (2), 177–203. https://doi.org/10.1080/01402382.2019.1702297

Caputo, R. (ed.) (2012) *Basic Income Guarantee and Politics: International Experiences and Perspectives on the Viability of Income Guarantee*. Springer.

Chandra Kruse, L., Bergener, K., Conboy, K., Lundström, J.E., Maedche, A., Sarker, S., Seeber, I., Stein, A. and Tømte, C.E. (2023) Understanding the digital companions of our future generation. *Communications of the Association for Information Systems* 52, 465–479. https://doi.org/10.17705/1CAIS.05218

Chen, L., Chen, P. and Lin, Z. (2020) Artificial intelligence in education: A review. *IEEE Access* 8, 75264–75278. https://doi.org/10.1109/ACCESS.2020.2988510

Chiu, T.K.F., Xia, Q., Zhou, X., Chai, C.S. and Cheng, M. (2022) Systematic literature review on opportunities, challenges, and future research recommendations of artificial intelligence in education. *Computers and Education: Artificial Intelligence* 4, 100118. https://doi.org/10.1016/j.caeai.2022.100118

Costa Dias, M., Joyce, R., Postel-Vinay, F. and Xu, X. (2020) The challenges for labour market policy during the Covid-19 pandemic. *Fiscal Studies* 41 (2), 371–382. https://doi.org/10.1111/1475-5890.12233

Deshpande, M. and Mueller-Smith, M. (2022) Does welfare prevent crime? The criminal justice outcomes of youth removed from SSI. *The Quarterly Journal of Economics* 137 (4), 2263–2307. https://doi.org/10.1093/qje/qjac017

Dimov, D. (2022) Conceptual model of automated trading systems implementation. *ROBONOMICS: The Journal of the Automated Economy* 3, 25. See https://journal.robonomics.science/index.php/rj/article/view/25 (accessed 20 December 2022).

Fan, R. and Cherry, M.J. (eds) (2021) *Sex Robots: Social Impact and the Future of Human Relations*. Springer Nature.

Flavián, C., Ibáñez-Sánchez, S. and Orús, C. (2019) The impact of virtual, augmented and mixed reality technologies on the customer experience. *Journal of Business Research* 100, 547–560. https://doi.org/10.1016/j.jbusres.2018.10.050

Fuchs, C. (2021) *Marxist Humanism and Communication Theory: Media, Communication and Society* (Vol. 1). Routledge.

Grahl, J. and Lysandrou, P. (2014) The European Commission's proposal for a financial transactions tax: A critical assessment. *JCMS: Journal of Common Market Studies* 52 (2), 234–249.

Guerreiro, J., Rebelo, S. and Teles, P. (2022) Should robots be taxed? *The Review of Economic Studies* 89 (1), 279–311. https://doi.org/10.1093/restud/rdab019

Gunkel, D.J. (2018a) *Robot Rights*. MIT Press.

Gunkel, D.J. (2018b) The other question: Can and should robots have rights? *Ethics and Information Technology* 20, 87–99. https://doi.org/10.1007/s10676-017-9442-4

Ismail, H.N., Iqbal, A. and Nasr, L. (2019) Employee engagement and job performance in Lebanon: The mediating role of creativity. *International Journal of Productivity and Performance Management* 68 (3), 506–523. https://doi.org/10.1108/IJPPM-02-2018-0052

Ivanov, S. (2017) Robonomics: Principles, benefits, challenges, solutions. *Yearbook of Varna University of Management* 10, 283–293.

Ivanov, S. (2021) Robonomics: The rise of the automated economy. *ROBONOMICS: The Journal of the Automated Economy* 1, 11. https://journal.robonomics.science/index.php/rj/article/view/11 (accessed 15 March 2021).

Ivanov, S. (2022) AI2AI marketing: Foundations and research agenda. *ROBONOMICS: The Journal of the Automated Economy* 3, 26. https://journal.robonomics.science/index.php/rj/article/view/26 (accessed 20 December 2022).

Jones, B.D. and Baumgartner, F.R. (2005) A model of choice for public policy. *Journal of Public Administration Research and Theory* 15 (3), 325–351. https://doi.org/10.1093/jopart/mui01

Khatib, O., Yeh, X., Brantner, G., Soe, B., Kim, B., Ganguly, S. ... and Creuze, V. (2016) Ocean one: A robotic avatar for oceanic discovery. *IEEE Robotics & Automation Magazine* 23 (4), 20–29. https://doi.org/10.1109/MRA.2016.2613281

Kovacev, R. (2020) A taxing dilemma: Robot taxes and the challenges of effective taxation of AI, automation and robotics in the fourth industrial revolution. *Ohio State Technology Law Journal* 16, 182–217.

McDonough, B. and Bustillos Morales, J. (2020) *Universal Basic Income*. Routledge.

Neuhofer, B., Buhalis, D. and Ladkin, A. (2014) A typology of technology-enhanced tourism experiences. *International Journal of Tourism Research* 16 (4), 340–350.

Organisation for Economic Co-operation and Development (OECD) (2017) *The Next Production Revolution: Implications for Governments and Business*. OECD Publishing.

Peeters, A. and Haselager, P. (2021) Designing virtuous sex robots. *International Journal of Social Robotics* 13 (1), 55–66. https://doi.org/10.1007/s12369-019-00592-1

Pereira, R. (ed.) (2017) *Financing Basic Income: Addressing the Cost Objection*. Springer.

Pereira, R. (ed.) (2023) *Financing Basic Income: A Dual Income Proposal*. Springer Nature.

Peters, M.A., Jandrić, P. and Means, A.J. (eds) (2019) *Education and Technological Unemployment*. Springer.

Preparing for The Future of Artificial Intelligence (2016, 2 October) Published by the Executive Office of the President of the USA, National Science and Technology Council and the Committee on Technology. See https://www.whitehouse.gov/sites/default/files/whitehouse_files/microsites/ostp/NSTC/preparing_for_the_future_of_ai.pdf (accessed 14 October 2016).

Schwitzgebel, E. (2023) The full rights dilemma for AI systems of debatable moral personhood. *ROBONOMICS: The Journal of the Automated Economy* 4, 32. See https://journal.robonomics.science/index.php/rj/article/view/32 (accessed 21 May 2023).

Sheahen, A. (2012) *Basic Income Guarantee: Your Right to Economic Security*. Palgrave Macmillan.

Stevens, Y.A. and Marchant, G.E. (2017) Policy solutions to technological unemployment. In K. LaGrandeur and J.J. Hughes (eds) (2017) *Surviving the Machine Age: Intelligent Technology and the Transformation of Human Work* (pp. 117–130). Palgrave Macmillan.

Straubhaar, T. (2017) On the economics of a universal basic income. *Intereconomics* 52 (2), 74–80.

The National Artificial Intelligence Research and Development Strategic Plan (2016, 2 October) Published by the Executive Office of the President of the USA, National Science and Technology Council and the Committee on Technology. https://www.whitehouse.gov/sites/default/files/whitehouse_files/microsites/ostp/NSTC/national_ai_rd_strategic_plan.pdf (accessed 14 October 2016).

Thoits, P.A. (2012) Role-identity salience, purpose and meaning in life, and well-being among volunteers. *Social Psychology Quarterly* 75 (4), 360–384.

Thuemmel, U. (2023) Optimal taxation of robots. *Journal of the European Economic Association* 21 (3), 1154–1190. https://doi.org/10.1093/jeea/jvac062

Tigard, D. (2023) On respect for robots. *ROBONOMICS: The Journal of the Automated Economy* 4, 37. https://journal.robonomics.science/index.php/rj/article/view/37 (accessed 27 June 2023).

Tobin, J. (1974/2015) *The New Economics One Decade Older*. Princeton University Press.

Velasco, C. and Obrist, M. (2020) *Multisensory Experiences: Where the Senses Meet Technology*. Oxford University Press.

World Economic Forum (WEF) (2023) The Future of Jobs Report 2023. World Economic Forum report. https://www.weforum.org/reports/the-future-of-jobs-report-2023 (accessed 18 June 2023).

Xie, X. and Wang, T. (2023) Artificial intelligence: A help or threat to contemporary education. Should students be forced to think and do their tasks independently? *Education and Information Technologies* 1–15. https://doi.org/10.1007/s10639-023-11947-7

Part 3
Robonomics and Future Tourism

4 Implications of Robonomics on Future Tourism: An Overview

Craig Webster and Stanislav Ivanov

4.1 Introduction

While the history of the future is not yet written, many drivers, including technological ones, will impact not just the way that the economy will function but will also have ramifications for the entire economy. Travel, tourism and hospitality will be increasingly automated, requiring a different set of expectations from the perspective of the consumer and a different service delivery system from the perspective of the service provider. As such, it will lead to a new way of doing things and here we lay out some of the components of the tourism ecology and how they will be transformed by robonomics.

The robonomic revolution will have a massive impact on economies and societies, as already elaborated on in Chapters 1–3. Since new technologies will be increasingly integrated into economies and societies in tangible and ubiquitous ways and will disrupt many of the more traditional ways of providing products and services to consumers, tourism, as a critical part of the service economy, will have to adjust to the new reality and new market. For example, in the future robonomic economy, a great deal of labour will be done using automation technologies. Tourism will cease to be 'people business' where people serve people and much of the service provision will be implemented by artificial autonomous agents, including service robots. Because of the massive advances in technologies, we will also see changes in markets and society, something in line with what has been noted for many years by philosophers of the future (see, e.g. Toffler, 1980).

Section 2 of the book focused on the theoretical fundamentals of robonomics as a future economic system based on artificial intelligence (AI), robots and other automation technologies while the use of human labour is minimised. This chapter discusses the general implications of robonomics for future tourism, while the following chapters provide deeper insights into some specific topics – the automated tourism/ hospitality company of the future, creating tourist experiences with automation technologies, designing robot-friendly hospitality facilities and human–robot relationships in tourist resorts.

4.2 Robonomics and Implications for Tourism

Robonomics will bring about significant changes in the way tourism in produced, consumed, experienced, valued and financed. Table 4.1 presents the characteristics, benefits, challenges and solutions to the challenges of robonomics and their respective implications for tourism.

There are a number of ways in which tourism will be challenged as an industry in the new robonomic system. First, the general nature and logic of the economic system will have evolved more deeply into the Fourth Industrial Revolution. For example, one of the key characteristics of the robonomic society will be the high level of automation in the provision of products and services, which implies that there will be high levels of productivity using capital. Such a revolution will lead to a major shift in the entire workforce (Tschang & Almirall, 2021). In terms of the tourism industry, this means that companies will employ fewer people and will have to be very selective about which humans they choose to work for them. However, it will also mean that tourism and hospitality employees will have to be highly skilled and understand the logic and capabilities of the automation technologies being employed in enterprises. In this way, not only will the size of the employed workforce be reduced but also the qualities of the workforce that remain employed will be qualitatively different from the workforce in operation prior to the Fourth Industrial Revolution. Already, there is a substantial digital divide in the workforce (Carlisle *et al.*, 2023), and with an increasingly digitalised economy a larger share of the workforce may be unable to keep pace with the technological demands of the workplace. One externality of this is that many humans may not be able to keep up and be employed by enterprises.

An additional employment-related general characteristic of robonomics is that autonomous agents will be making many of the decisions in enterprises. While this relieves humans of the burden of many decisions, it also means that humans will have to ensure that autonomous agents are monitored to make certain that decisions are made in accordance with what is permitted and desired (Ivanov, 2023). Moreover, granting economic rights to robots and sophisticated artificial autonomous agents (Gunkel, 2018) will lead to the appearance of completely automated tourism and hospitality companies that are not only run by but also owned by autonomous agents. Potential tourists will also own artificial autonomous agents that will help them plan, organise, book, pay for and experience their trips. This means that robots and other artificial autonomous agents belonging to tourists need to be treated as customers because their actions are indistinguishable from the actions of human tourists (Ivanov & Webster, 2017b). Therefore, robots and other

Table 4.1 Robonomics and implications for tourism

Aspects of robonomics	Implications for tourism
Characteristics of robonomics	
• High level of automation	• Capital-intensive provision of goods and services • Robots and artificial autonomous agents provide tourism/hospitality services • Robot-friendliness of hospitality facilities
• Artificial autonomous agents serve as economic agents	• Reduced need for human decision-makers • Travel decisions taken by AI rather than by humans • Tourist companies focus on AI2AI marketing • Automated tourism and hospitality companies owned and operated by artificial autonomous agents
• Few humans work	• Need for a small and highly skilled labour force • Human labour largely used to correct for service failures and tasks that are difficult to automate • Decrease in tourism employment • Humans have sufficient time for tourism activities
• Disconnection between employment and incomes	• Tourism is not a significant source of income
• Knowledge and creativity are sources of competitive advantage not capital and labour abundance	• The increased need for a creative workforce to design new and revolutionary products to compete effectively in the market
• Überveillance, privacy and behavioural control	• Service providers will have data to create very personalised services • Most human activities will become monetised data
Benefits of robonomics	
• Decrease in costs and prices	• Changes in the profitability for high-tech service provisions • Tourism and hospitality services are affordable
• Improved environmental sustainability of operations	• Sustainable tourism is the norm • Increased capacity possible even with sustainability goals
• Accelerated scientific research	• Rapid incorporation of new technologies into operations • Shortened tourism/hospitality product life cycle
• Improved quality of life	• Changes in the products offered to customers – focus on well-being
• Global government and global citizenship	• Increased standardisation of regulations • No visas required for travel • No currency exchange risk due to global currency
• Accelerated space exploration	• Increased opportunities for space tourism • Normalisation of space tourism
Challenges of robonomics	
• Unemployment and relative overpopulation	• There will be governmental pressure to hire people, including in state-provided tourism/leisure activities, while there will be a decreasing need for human labour in enterprises
• Lower employment incomes	• Tourism employees may still be underpaid compared to employees in other sectors
• Psychological problems of people	• Tourism focus on well-being • Facilities modified to cater to tourists with psychological issues

(Continued)

Table 4.1 (Continued)

Aspects of robonomics	Implications for tourism
• Political instability	• Inability to make long-term plans based on short-term and mid-term political turbulence • Hospitality facilities might be destroyed during unrest • Hospitality design may incorporate greater security measures to protect customers and employees
• Functional illiteracy	• Smaller pool of potential employees who have the necessary skills to work in tourism • Increased importance of ease of use of technological products implemented in tourism and hospitality companies • Evolution away from written signage and instructions towards audio and icons
• Social divisions	• Two large tourism market segments (technology-delivered low-cost services and human-delivered high-priced services) with various shades of grey in between them • Branding of services towards different markets based on social divisions
• Changes in human values	• Industry will have to tailor products and services in line with the changes in values • Marketing to focus on the emerging values of various segments of the market
• Possible decrease in population in the long term	• The need to utilise long-term market decline as a premise in strategic planning
Solutions to the challenges of robonomics	
• Mandating employment	• Tourist companies forced to hire employees to meet government requirements
• Government job creation	• Some job opportunities in government job creation projects (e.g. construction of airports and highways)
• Work sharing	• Increased part-time employment in tourism • Increased failures in customer service due to the dilution of responsibilities in the workplace
• Employment impact statements	• Pressures to illustrate that employment in tourism enterprises has an externality that is a social benefit
• Tax policies	• Changes in the way tourism and hospitality companies are run to avoid taxation that would undermine profitability
• Financial incentives for job creation	• Creation of tourism and hospitality jobs that would otherwise be redundant via technological unemployment
• Constant, fluid and free lifelong education	• Continuous training opportunities for tourism and hospitality employees • Additional costs for enterprises to create or purchase training programmes
• Entertainment, tourism and leisure activities	• The creation of state-subsidised entertainment incorporated into tourism to encourage social engineering conducive to economic stability • Increased political importance of tourism to keep society under control and avoid political tensions and unrest
• Volunteering	• The importance of creating volunteer opportunities in tourism to support morale in the population conducive to public service • Incorporation of initiatives in tourism-related enterprises to encourage volunteerism among employees

(Continued)

Table 4.1 (Continued)

Aspects of robonomics	Implications for tourism
• Universal basic income	• People have the financial means to travel • A large segment of the market includes people who have income only from UBI • Many new start-ups, including in tourism and hospitality, because UBI serves as a social safety net
• Robot-based taxation	• Tourism and hospitality companies invest in specific technologies in ways that can avoid taxation that would otherwise undermine profitability
• Birth control/birth right taxes	• The need to plan for a sustained population reduction and changes to the market dependent on that model • Ageing customer base and decreased service availability for large families
• Redefinition of human rights	• Changes in demography and the purchasing power of people based on the specific composition of people's rights
• Robot/AI rights	• Serving robots as tourists • Automated tourism and hospitality companies owned and operated by artificial autonomous agents

artificial autonomous agents in tourism and hospitality companies will need to target the robots and other artificial autonomous agents belonging to potential human customers, i.e. they will need to apply AI2AI marketing (Ivanov, 2022), while robots themselves might become tourists (Ivanov, 2019a). Meanwhile, the massive use of robots in tourism will force architects to design robot-friendly facilities to allow robots to fulfil their tasks (Ivanov & Webster, 2017a).

Apart from employment-related issues, creativity and access to data will offer competitive advantages to enterprises. For tourism, it means that companies will have to come up with not only increasingly interesting and creative products for the market but also that such interesting and creative products are based on market intelligence that may be quite specific to individuals and markets. This also implies that the supply side of the equation will have to have access to data, analytical capabilities and creative solutions in order to ensure that enterprises can continue to produce tourism and hospitality services that are in demand for the changing market, although some risks are inherent when such access to data becomes available (Yallop & Seraphin, 2020). Part and parcel of this is that the robonomic world will have data available at the individual level, something that may seem invasive but for tourism and hospitality companies it will ensure substantial market insights to understand and predict consumer behaviour to create highly personalised services for individual customers that will work to ensure that many customers are return customers (Silva *et al.*, 2020).

With robonomics, there are definitive benefits that will give some advantages to the tourism industry. For example, increases in productivity should lead to a long-term reduction in costs and prices for consumers, following the initial investment costs of automation. In addition, there could be improved environmental benefits, since automation technologies will increase economic efficiency and save resources, although automation may not always have a linear and positive impact on environmental concerns (Yang et al., 2023). The acceleration of scientific research should provide opportunities for additional labour-saving and cost-saving technologies in tourism enterprises. The improved quality of life should also offer opportunities for customers, as we can expect older and healthier populations to be customers at tourism facilities (Webster & Ivanov, 2023). In addition, globalised regulatory capabilities will increase the standardisation of regulations so that enterprises will have a simplified operational structure, since there will be a standard and global regulatory framework and a global currency unit. Finally, increased space exploration will enable the tourism market to further expand beyond terrestrial limitations – e.g. space tourism in hotels orbiting the Earth, on the Moon or on Mars (Toivonen, 2021).

While robonomics creates many opportunities, there will also be challenges, mostly because of the social transformation from the institutions of the old economy. One of the key issues is that many people may not have the skills that are needed in the new economy; indeed, it may even be the case that the new economy will make people less capable intellectually, since automation may undermine the cognitive capabilities of people (Breton & Bossé, 2003). There will also be political pressures for various tourism companies to hire people when individuals do not have the skills for the available jobs and the industry does not need them. While such Keynesian make-work projects may have short-term benefits, they will come at a cost. During the transition to the mostly robonomic economy, there will be political turbulence with vast changes in the job market and shifts in attitudes, and tourism-related enterprises will have to make adjustments for this, especially in strategic planning. There will also be a very different population (not just ageing) with younger people functionally illiterate, enhanced social divisions, changes in values and long-term population decline (Webster, 2021). People employed will require a great deal of continuous training and will likely need lifelong education. Fortunately, much of the provision of education and training will also be provided by more automated education and training systems (Ivanov & Soliman, 2023). The tourism industry will have to look at the new cleavages and changes and react appropriately, creating new leisure opportunities and products consistent with the new reality. For example, with changes in values systems, tourism enterprises may develop more varied products that provide for specific markets, such as volunteer tourism opportunities or increased interactive tourism opportunities in historical preservation for cultural tourism. With massive social changes stemming from a very different economic system, cultural values may shift beyond post-materialist values systems that have been observed in recent decades (Inglehart, 1990) and the market for tourism and hospitality services will have to adapt products to the new system of values held by new consumers.

While the technologies that are increasingly empowered will play a major role in the economy, there will be pressures from governmental authorities to deal with the externalities and the tourism industry will have to expect increased pressure to comply with governmental demands and changes in policies. For example, with a decline in the need for human labour, it is expected that the government will create jobs or mandate hiring for various industries. Tourism enterprises will have to cooperate with the government on many schemes, and some of them may be beneficial if accompanied by subsidies, although they are not needed for practical reasons. A social and possibly also political change may be the expansion of job sharing in tourism, since less labour may be needed in workplaces, and individuals will be less dependent on salaries because they will receive a universal basic income (UBI). Such job sharing is likely to have a negative impact on the provision of services in service failure environments, although it is expected that the systems will be so automated that such failures will be few and far between.

As a social innovation, UBI would ensure the purchasing power of people regardless of their employment status, although there are some concerns that the implementation of such a scheme comes with some risks (Fouksman & Klein, 2019). The idea is that UBI will provide the financial means for people to pay for tourism services because their incomes from other sources (e.g. employment) might not be sufficient or exist at all. The safety net function of UBI may stimulate many people to start entrepreneurial activities in tourism and hospitality and the establishment of many start-ups. At the same time, the levelling effect of UBI may divide the tourism and hospitality markets into two large groups with more than 50 shades of grey between them. At the one extreme will be customers who will rely only on their UBI and may afford to pay for cheap technology-delivered tourism and hospitality services while the second group may afford to pay for expensive human-delivered services (Ivanov, 2019b). A bifurcated market will create market opportunities for entrepreneurs who want to provide services to the majority who will be dependent on UBI.

In a robonomic society, many people will have a great deal of time and energy and will not be distracted with what humans have been distracted with for millennia, the struggle to survive. Because of this, there will be increased concern about the growth of psychological problems, some of which can be remedied with distractions, such as entertainment, tourism and leisure activities. Such issues may also be ameliorated with the provision of volunteering opportunities for people, giving individuals a sense of purpose in society. Hence, tourism and especially volunteer tourism will play an important role in maintaining the fabric of society.

In addition, entertainment, tourism and leisure will become critical aspects of maintaining social order, the modern equivalent of 'bread and circus' that was used to maintain order in ancient Rome. In that way, it is expected that tourism-related activities will be used elites' intention to maintain social and political stability among the population.

4.3 The Role of Tourism in a Robonomic World

In a robonomic world, the economy will be greatly transformed from what many people in 1968 would have understood. Most importantly and most obviously, the vast majority of goods and services will be produced and delivered using very high levels of automation. In this robonomic economy, the use of automation will have moved a great deal beyond the mechanical reproduction of goods, as typically produced in highly automated factories. The externalities of this will be social and economic in nature. Most obviously, fewer people will be needed in order to create value in the economy, meaning that in many ways, humans will largely shift their value to the economy from creators of value to consumers of value. Another major externality verges on the spiritual, the search for a purpose in life since humans will largely not be needed to create value in the economy. Despite the many ways in which the robonomic world will produce vast amounts of wealth in efficient ways, the externality of finding ways that humans can be part and parcel of the economy and society will have to be developed.

Because of the changes and the externalities of robonomics, tourism not only has challenges but also has a very real and utilitarian purpose. Tourism is one of the human practices that illustrates an elevated quality of life in a population. Since the world will be shifting to a radically different type of economy and society, the affordable provision of quality leisure services, of which tourism is a key one, is critical to illustrate to the population that the robonomic way of doing things is a benefit for nearly everyone, even those who have suffered from technological unemployment. To promote social, political and (ultimately) economic stability, tourism has to flourish with quality products to offer to the masses. Fortunately, automation technologies can decrease operational costs and increase the quality of tourism products (Belanche *et al.*, 2021).

Tourism can be used as a general tool to keep people entertained, psychologically happy and engaged in society. While the Roman Empire used bread and circuses to help maintain stability, one component of the modern equivalent may be tourism and events (Waitt, 2008). Tourism can be used as a political tool to promote social and political stability. However, it is not just a distraction but it also offers ways to engage people socially. For example, volunteer tourism can create opportunities for people to feel engaged and provide a sense of purpose in their society. This would also work in ways to ensure that the psychological outcomes of modern society are addressed/mitigated, since giving individuals a sense of purpose and social connections is something that tourism, as an activity, can do.

4.4 Conclusion

As sociobiologist Edward O. Wilson stated, 'The real problem of humanity is the following: We have paleolithic emotions, medieval institutions and godlike technology' (quoted in Harris, 2019). The transition to robonomics and the turbulence of the process is not just a transition from one way of doing things but also includes the need to socially engineer a population to the new reality, after having a population that has been socialised into institutions that were formed in a different era. First and foremost, to be successful, the transformed tourism industry will have to be designed to meet the paleolithic emotions of the human. Future researchers will have to work closely to fully understand the psychology of humans to ensure that what is marketed to humans will fit their sometimes unstated and subconscious wants and needs. Tourism, as a vital component of the new economy, will play a critical role in maintaining social order, giving people a sense of purpose and providing a vast array of services. Because society will change so drastically due to changes in the economic base, tourism and its related industries will have to metamorphose into something engaging, useful and supportive of the population while being mindful of the Palaeolithic brain.

In the new reality, nearly all industries will be highly automated, few humans work, while there will be the practical ability to make an abundance of goods and services in an efficient and effective way; a truly unique historical accomplishment for humanity. While the Palaeolithic brain of the human may remain the same, the institutions in which humans function will have to change. The post-materialist values systems (see Inglehart, 1990) will have to be taken into account in the formation of new institutions on which tourism will be based and function. Tourism will need to address a large population dependent on UBI and a less literate population, albeit a population that demands goods and services that it has not had to work for to attain. The ethos of new customers will have to lead to new visions of what tourism and hospitality are to be.

The following chapters explore various aspects of robonomics on society and the role of tourism in the new socioeconomic milieu. They explore the general characteristics of the robonomic society, delve deeply into the benefits of robonomics and discuss the challenges and solutions to problems in a robonomic society. While we are still in the first part of this Fourth Industrial Revolution, we have the luxury of contemplating the major issues that we will face and discuss intelligent and dynamic solutions to the problems humanity will confront in this brave new world. It is with hope that humans can create intelligent institutional responses to navigate the problem of integrating their Palaeolithic emotions with the godlike technology that we humans have created. Tourism is one of the key service sectors in which the robonomic world will engage citizens and consumers to elevate their quality of life and work in ways to ensure social and economic stability. The following chapters explore issues related to its role in the new reality of tourism.

References

Belanche, D., Casaló, L.V. and Flavián, C. (2021) Frontline robots in tourism and hospitality: Service enhancement or cost reduction?. *Electronic Markets* 31, 477–492. https://doi.org/10.1007/s12525-020-00432-5

Breton, R. and Bossé, É. (2003) The cognitive costs and benefits of automation. Paper presented at the RTO HFM Symposium on 'The Role of Humans in Intelligent and Automated Systems', Warsaw, Poland, 7–9 October, and published in RTO-MP-088. See https://www.sto.nato.int/publications/STO%20Meeting%20Proceedings/RTO-MP-088/MP-088-01.pdf (accessed 31 October 2023).

Carlisle, S., Ivanov, S. and Dijkmans, C. (2023) The digital skills divide: Evidence from the European tourism industry. *Journal of Tourism Futures* 9 (2), 240–266. https://doi.org/10.1108/JTF-07-2020-0114

Fouksman, E. and Klein, E. (2019) Radical transformation or technological intervention? Two paths for universal basic income. *World Development* 122, 492–500. https://doi.org/10.1016/j.worlddev.2019.06.013

Gunkel, D.J. (2018) *Robot Rights*. MIT Press.

Harris, T. (2019) Our brains are no match for our technology. *The New York Times*. https://www.nytimes.com/2019/12/05/opinion/digital-technology-brain.html (accessed 31 October 2023).

Inglehart, R. (1990) *Culture Shift in Advanced Industrial Society*. Princeton University Press. https://doi.org/10.2307/j.ctv346rbz

Ivanov, S. (2019a) Tourism beyond humans: Robots, pets and teddy bears. In G. Rafailova and S. Marinov (eds) *Tourism and Intercultural Communication and Innovations* (pp. 12–30). Cambridge Scholars Publishing.

Ivanov, S. (2019b) Ultimate transformation: How will automation technologies disrupt the travel, tourism and hospitality industries? *Zeitschrift für Tourismuswissenschaft* 11 (1), 25–43.

Ivanov, S. (2022) AI2AI marketing: Foundations and research agenda. *ROBONOMICS: The Journal of the Automated Economy* 3, 26. https://journal.robonomics.science/index.php/rj/article/view/26 (accessed 19 December 2022).

Ivanov, S. (2023) Automated decision-making. *Foresight* 25 (1), 4–19, https://doi.org/10.1108/FS-09-2021-0183

Ivanov, S. and Webster, C. (2017a) Designing robot-friendly hospitality facilities. *Proceedings of the Scientific Conference 'Tourism. Innovations. Strategies'*, 13–14 October 2017, Bourgas, Bulgaria, pp. 74–81.

Ivanov, S. and Webster, C. (2017b) The robot as a consumer: A research agenda. *Proceedings of the 'Marketing: Experience and Perspectives' Conference*, 29–30 June 2017, Science and Economics, University of Economics-Varna, Bulgaria, pp. 71–79.

Ivanov, S. and Soliman, M. (2023) Game of algorithms: ChatGPT implications for the future of tourism education and research. *Journal of Tourism Futures* 9 (2), 214–221. https://doi.org/10.1108/JTF-02-2023-0038

Silva, E.S., Hassani, H. and Madsen, D.Ø. (2020) Big Data in fashion: Transforming the retail sector. *Journal of Business Strategy* 41 (4), 21–27. https://doi.org/10.1108/JBS-04-2019-0062

Toffler, A. (1980) *The Third Wave*. Morrow.

Toivonen, A. (2021) *Sustainable Space Tourism: An Introduction*. Channel View Publications.

Tschang, F.T. and Almirall, E. (2021) Artificial intelligence as augmenting automation: Implications for employment. *Academy of Management Perspectives* 35 (4), 642–659. https://doi.org/10.5465/amp.2019.0062

Waitt, G. (2008) Urban festivals: Geographies of hype, helplessness and hope. *Geography Compass* 2, 513–537. https://doi.org/10.1111/j.1749-8198.2007.00089.x

Webster, C. (2021) Demography as a driver of robonomics. *ROBONOMICS: The Journal of the Automated Economy* 1, 12. https://journal.robonomics.science/index.php/rj/article/view/12

Webster, C. and Ivanov, S. (2023) Robots, artificial intelligence and service automation in tourism and quality of life. In M. Uysal and M.J. Sirgy (eds) *Handbook of Tourism and Quality-of-Life Research II. International Handbooks of Quality-of-Life* (pp. 533–544). Springer. https://doi.org/10.1007/978-3-031-31513-8_36

Yallop, A. and Seraphin, H. (2020) Big data and analytics in tourism and hospitality: Opportunities and risks. *Journal of Tourism Futures* 6 (3), 257–262. https://doi.org/10.1108/JTF-10-2019-0108

Yang, X., Luan, F., Zhang, J. and Zhang, Z. (2023) Testing for quadratic impact of industrial robots on environmental performance and reaction to green technology and environmental cost. *Environmental Science and Pollution Research* 30, 92782–92800. https://doi.org/10.1007/s11356-023-28864-4

5 Future Trajectories for Automated Tourism and Hospitality Services

Ellis Urquhart

5.1 Introduction

Automation in tourism and hospitality is not a new phenomenon; however, its proliferation and rapid enhancement have garnered increasing academic attention (Tussyadiah, 2020). Innovations such as self-service platforms (Wei *et al.*, 2017), service robotics (Tung & Law, 2017) and artificial intelligence adoption (Buhalis *et al.*, 2019) have already begun to fundamentally alter the traditional service encounter and shift consumers into 'prosumers' (Ritzer, 2015). However, despite these practices existing within human-to-human environments, increasing attention is being paid to the future trajectory of the tourism and hospitality industry. Namely, scholars within future-orientated research are looking towards the next generation of tourism experiences that feature enhanced automation (Ivanov & Webster, 2019). Arguably, much of the motivation for adopting automation is in response to changing demography, demand and the availability of labour. The apparent population decline in developed countries alongside the increased technological capabilities of emerging generations creates significant opportunities for automation within tourism (Webster, 2021; Webster & Ivanov, 2020). As noted by Wang *et al.* (2016), the increasing preference for self-service, autonomy and personalisation among new digital tourists will stimulate the industry into more innovative future practices of 'smart tourism'. Conversely, there are equally valid fears associated with automation particularly in relation to perceived job security, career trajectory and a wider fear of being replaced within the workplace (Ivanov *et al.*, 2020b). While the current desire for synchronous communication and response is only likely to continue into the future (Buhalis & Sinarta, 2019; Loureiro, 2017), the development of automation not only provides opportunities but also raises questions about human–machine interaction (Naumov, 2019). This is compounded by a growing skills shortage within the tourism sector with recent UK research indicating an increasing density of vacancies particularly in labour-intensive roles that could be filled by autonomous technologies (Economic Insight, 2019). The purpose of this chapter is to consider the future trajectories for a range of tourism and hospitality businesses and to explore the potential challenges and issues of future automation practices.

5.2 Multi-Level Automation

As Bolton *et al.* (2018) discussed, automation and artificial intelligence pose several opportunities for organisations in relation to large-scale data collection and automating routine tasks. Similarly, the potential to reduce human error through automated, standardised practices and the efficiency benefits of accelerated service provision provide lucrative prospects for tourism and hospitality providers. However, costs to tourism organisations remain a key challenge. As noted by Wirtz *et al.* (2018), current robotic technology manufacturing, maintenance and design are accompanied by significant financial costs; however, production costs into the future are likely to fall through mass production. As such, it is important to consider the various levels at which automation can occur within organisations before exploring sector-specific applications. As Figure 5.1 shows, service automation can and will continue to be applied at three organisational levels. At a business level, automated platforms can be used strategically for big data collection and to more efficiently match consumer demand with resource allocation. From an employee perspective, automation provides opportunities for staff-led platforms that seamlessly integrate front- and back-office operations for an enhanced personalisation experience. Finally, at the consumer level, automation has the potential to streamline the collection of service information, transactions and the post-experience journey through targeted customer relationship management and intuitive communication. From

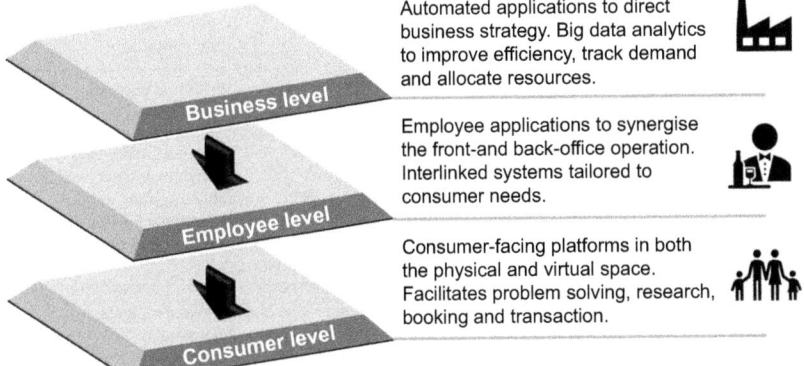

Figure 5.1 Multi-level automation in tourism and hospitality businesses (Source: Author)

a futures perspective, Gretzel *et al.* (2020) highlight the particularly vital role that technological innovation will play within the tourism industry for future generations. The authors highlight the industry's recent response to COVID-19 as a major catalyst to rethink and reframe the nature of technological mediation in the tourism, hospitality and leisure sectors. With an increased focus on fluctuating demand, rapid communications, augmented experiences and fragile resource allocation – smart tourism and automated systems will become even more critical for the industry.

5.3 Future Trajectories for Automated Tourism and Hospitality Services

Gretzel *et al.* (2015) argued that further insight into the future role of technology in tourism consumption (access, experience, perception of value), service provision (data, capacity, market, business models) and facilitation (governance, infrastructure, cost/benefit analysis) is needed within academic research. As Figure 5.2 shows, a variety of automated applications can be adopted in the tourism and hospitality businesses; however, these may be uniquely applied in transport, accommodation, attractions and food and beverage environments. The following sections explore the application of such technologies in these contexts through

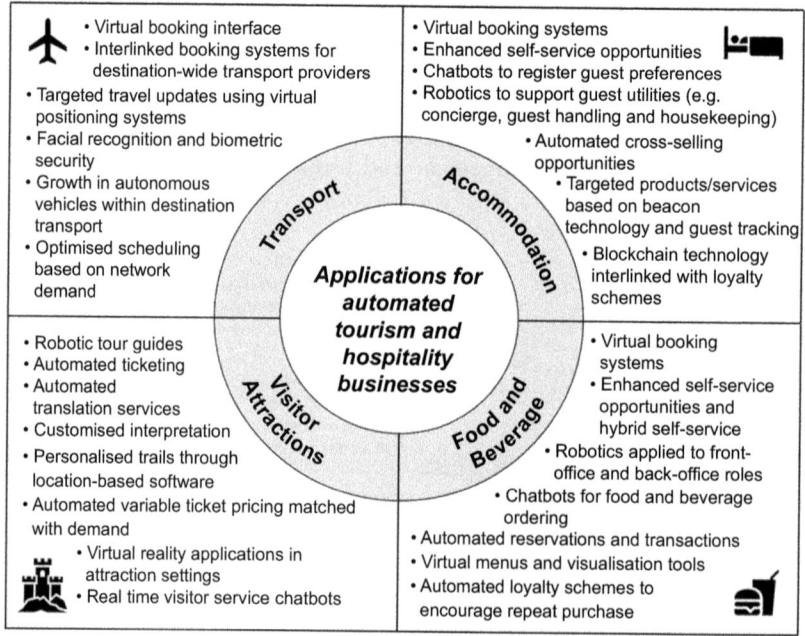

Figure 5.2 Applications for automated hospitality and tourism businesses (Source: Author)

a future-orientated lens. These future trajectories assume a mass acceptance approach to automated technology whereby businesses actively invest, adopt and innovate to firmly embed technologies into tourism and hospitality operating models (Urquhart, 2019). The holistic challenges and opportunities afforded by this approach are then explored in further depth.

5.3.1 Transport

As a sector reliant on efficiency and connectivity, passenger transportation is in prime position to develop automated capabilities. The future movement towards 'smart cities', which encompass urban landscapes that are integrated with autonomous, integrated and shared technologies, will become increasingly embedded in destination development (Nikitas et al., 2020). For transport providers, future developments in existing automated technology act as a cornerstone to achieve seamless transport networks within interconnected systems. As suggested by Keuchel (2020), automation can be seen as a central component in future transport journeys, with digital maps and route networks increasingly accessible to consumers. The integration of voice activation and targeted notifications is likely to become further embedded in transport booking systems, offering intuitive and reactive guidance for passengers.

Automation and robotics also provide future avenues for personalising the passenger experience. Previous examples can be illustrated through the launch of 'Pepper' – the humanoid robots temporarily deployed at Eurostar's St Pancras Station in London and Taipei International Airport to provide passenger services, directions and guidance (Reid, 2018). Transport operators of the future may further capitalise on such technology to replace current service roles such as passenger handling, processing and logistics to increase efficiency and reduce human error. Virtual bots, holograms and digital kiosks in place of human-operated information desks are likely to become a feature of the future servicescape with technologies capable of providing autonomous information with which consumers can independently engage (Manthiou et al., 2021). Similarly, the potential for automated transport security is rapidly growing. With automated e-gates, biometric security and passenger screening developing in the aviation sector, future developments will likely make these more mainstream and increasingly the default mediator in international travel. The reliability of these platforms will undoubtedly increase over time. While Papadimitriou et al. (2020) argue that automation cannot replace the training and experience of aviation professionals, future platforms will inevitably become more alert to threats and inconsistencies in passenger security. Equally, automated platforms do not suffer from human limitations such as fatigue, failing health and ageing, providing opportunities for sustainable workforces in high-pressure, labour-intensive roles within international transport.

In line with the future focus on smart cities and interconnectivity, autonomous vehicles provide significant opportunities for transport providers. While still in its infancy, self-driving and semi-autonomous vehicles can already be found in urban infrastructures (such as Germany's Metro system and the MTR in Hong Kong). However, future opportunities for autonomous vehicles are diverse. As suggested by Hansen *et al.* (2016), in the case of rail transport, significant benefits in relation to increased capacity and reliability, energy savings and reduced operating costs provide lucrative drivers for further automating passenger services. Furthermore, Prideaux and Yin (2019) highlight the mobility and accessibility opportunities afforded by the removal of human drivers. Conversely, the mass introduction of autonomous vehicles would undoubtedly pose implications for transport management and employability. Long-standing careers in haulage and passenger transport (such as taxis, buses, monorail operators and chauffers) may become obsolete. This brings particular considerations for the removal of professions in the future destination infrastructure and the resulting socio-economic implications that would emerge.

Several sociopolitical issues may also arise from the mass introduction of autonomous vehicles. Future transport providers may soon have the ability to remove legislation such as age restrictions for drivers/passengers in an autonomous vehicle to accommodate diverse passenger profiles. Similarly, as suggested by Cohen and Hopkins (2019), autonomous vehicles may create changes in the future nighttime economy by reducing the likelihood of crime, anti-social behaviour, drunk driving, accidents and fatalities by removing human drivers involved in late-night transport. Similarly, the accessibility agenda, which will become even more critical in an ageing society, can be met with enhanced automated provision for disabled passengers or those with limited sensory perception. Wider social and environmental trends in the future (such as eradicating traffic congestion and sole occupancy vehicles) can also be recognised with autonomous peer-to-peer transport options, such as driverless car sharing (Prideaux & Yin, 2019). This provides further opportunities for reconfiguring automobiles (of varying scales) beyond transport and into stand-alone social spaces where elements of leisure, hospitality and events can be integrated within autonomous vehicles, strengthening the link between mobility and leisure experience.

From a passenger perspective, the future growth in autonomous vehicles may encounter challenges. As noted by Panagiotopoulos and Dimitrakopoulos (2018), the development and maintenance of consumer trust in driverless vehicles is a critical challenge for transport operators. While that may represent the current perception, businesses of the future will need to strategically consider how best to alleviate consumers' concerns and maintain safety. The advancement of intuitive and ubiquitous technology will likely address this issue, while also reaffirming the passenger benefits posed by autonomous transport such as increased comfort, connectivity and efficiencies in time and cost (Zhang *et al.*, 2019). However, the future growth of autonomous vehicles could bring unintended consequences to the consumer experience. The lack of manual control in next-generation vehicles may prohibit passengers from travelling during periods of malfunction, servicing or maintenance. Similarly, safety concerns associated with a collision in vehicles without manual failsafes may be considerable. Finally, in the future digital landscape, data security and integrity are more likely to be targets for cybercrime or manipulation. The threat of vehicle hacking, overriding or security breaches would continue to be a key area of concern for autonomous vehicles of the future.

5.3.2 Accommodation

The accommodation sector finds itself in a complex position with regard to automation. As suggested by Balasubramanian and Ragavan (2019), while there are significant opportunities for advanced technological adoption in hospitality, there is a need to retain the 'human heart' of the sector and the unique host–guest relationship. From this perspective, automation should be used to enhance value propositions and 'wow' factors for guests of the future. Conversely, Ristova and Dimitrov (2019) argue that the digitisation of hospitality is inevitable and indeed necessary to meet heightening guest expectations for targeted, active and efficient products/services. It is also highlighted that the accommodation sector is diverse to cater for a multitude of consumer demands. For example, fully automated hotels already exist in Asia and market themselves for the privacy and anonymity that guests will encounter – notably in relation to the 'love hotels' in Japan, Korea and Taiwan (Kim & Kim, 2018). Future trajectories for the accommodation sector may extend to this level of complete automation, but may equally apply digital tools to enhance existing offerings. Accommodation providers of the future will need to react not only to the changing nature of the workforce but also to changing market demand. As such, future consumers in certain markets (e.g. luxury bespoke experiences) may continue to value the social contact and relational quality offered by human personnel. Future service design planning will, therefore, need to focus on market segmentation to inform decisions about automated technology integration but creatively apply digital tools to enhance existing offerings.

As discussed by Seal (2019), automated technologies are already becoming embedded in accommodation providers. The growth of automated booking systems, chatbots and beacon-enabled push notifications is slowly becoming more common in international hotel chains. Future developments are expected to see the adoption of more of these technologies in small to medium-sized accommodation providers as infrastructure costs continue to reduce. Furthermore, many of the current automation systems are reliant on tangible smartphone access, whereas the future trajectory is most likely to utilise hands-free interfaces such as voice-activated speakers and wearable technologies to engage in more hands-free activity. The *in situ* guest experience is also likely to be enhanced through automated technologies. As suggested by Tuominen and Ascenção (2016), hotels of the future may explore 'polymorphic spaces' in which ambient room conditions (sound, scent and atmospherics) can be adjusted by individual guests at will. Although just one scenario, the concept of guest customisation using advanced technology can be seen as a significant growth area for the future of the sector (Kim & Han, 2020).

Academic and practitioner interest is growing in the adoption of robotics in the accommodation sector (Bowen & Morosan, 2018; Ivanov *et al.*, 2020a; Kuo *et al.*, 2017; Xu *et al.*, 2020). There is significant potential for robotic platforms to adopt certain roles within accommodation to maintain guest utilities (e.g. housekeeping), to maintain consistency and to standardise quality levels. Conversely, Kuo *et al.* (2017) suggest that the long-term trajectory for robotics in hospitality may help mitigate a growing skills shortage in certain future accommodation services. As such, robotics may be seen as not only 'role replacers' but also 'role creators' in the wider supply chain through the need for talented individuals involved in the design, research and development of hospitality robotics. Nevertheless, the application of robotics within accommodation settings in the future is uncertain. Baum (2019) suggests that automated platforms that replace front-line personnel may continue to prevail in the budget accommodation sector where the human-service dimension is perceived as less important. The key split for hospitality of the future will potentially relate to the presence of visible versus invisible robotics and automated systems. In the luxury sector, for instance, future operators may elect to utilise automation 'behind the scenes' while retaining authentic human contact as part of their differentiating strategy. Moreover, the adoption of robotics within hotels and accommodation providers will require the reconfiguration of hospitality spaces. As discussed by Ivanov and Webster (2017), future providers will need to consider robot-friendly facilities including access, repair/maintenance facilities, power management, navigation and sensor-based recognition technology to enable intelligent robots to fulfil tasks effectively.

5.3.3 Visitor attractions

Visitor attractions represent a broad sector of the tourism industry and can encompass heritage, cultural and entertainment-based experiences. The diversity of attractions within destinations leads to a number of opportunities for automated technologies in line with their specific management challenges. As such, Wang *et al.* (2016) suggest that 'smart' attractions of the future will focus on generating rich, real-time data about visitor behaviours, demands and expectations that attractions can then respond to within their product offering.

The use of automation for visitor management, while currently at an early stage, has further potential for attractions. E-ticketing has already become commonplace within the visitor attraction domain with museums, theme parks, heritage sites and visitor centres routinely offering web-based platforms or mobile applications to pre-book tickets (Guo *et al.*, 2014). In the future, however, there is likely to be a greater focus on the integration and intuitive packaging of ticketing options. As such, variable pricing can be manipulated into personalised attraction tours that incorporate a number of sites within one platform. Further advances in data analytics are likely to also enhance personalised recommendations for attractions within destinations that exist on shared booking systems. Prototypes of such an approach are functioning in Amsterdam with the 'I-Amsterdam' initiative that makes use of smartphone technology and personalised dashboards to allow visitors to access multiple venues, receive targeted visitor information based on purchase history and create tailored trails based on location tracking. However, in the future, interconnectivity between visitor attractions, leisure and cultural venues across a shared visitor management platform may provide lucrative avenues for visitor management.

Automations related to purchases are also a significant growth area. A notable current example is the use of MagicBands at Walt Disney World, Orlando, which employs wearable radio-frequency identification (RFID)-enabled wristbands gaining the wearer access to rides, attractions, events, rooms and souvenir purchases based on pre-set user permissions (Borkowski *et al.*, 2016). In extending this into the future, this personalised technology that seamlessly connects the visitor to attraction services could extend beyond wearable platforms. For example, attractions may be able to further adopt biometric controls (such as eye-scanning or fingerprint purchasing) to achieve similar goals. Conversely, future innovations may move to the implantation of nanotechnology into the human physiology, removing the need for tangible devices for purchasing, communicating or data capture (e.g. recording and image storage). While this may be significantly away from public use, examples of implanted technology exist in contemporary medical treatment and it is conceivable that innovations for consumer use may follow.

Further visitor management automations could build on existing flow monitoring and crowd handling techniques using big data. In future attractions, it is anticipated that greater emphasis will be placed on measuring and tracking visitor movements. As suggested by Mygind and Bentsen (2017), sensor-based tracking systems offer a number of opportunities for the attraction sector to analyse visitor preferences, flow and potential overcrowding points. While these can already be found in some large-scale museum settings, the future trajectory for this automated technology is profound. Attractions of various scales and sorts may invest further in these observation technologies to comprehensively understand their visitors' behaviour. This is particularly relevant in the heritage and natural sector where visitors engage with vast spaces and potentially fragile core resources. Additional understanding of visitor movements and behaviour provides critical management data to mitigate pressure and damage in sensitive locations. Equipped with this data, attractions of the future would be able to concentrate visitors onto low-impact routes throughout attractions and better diffuse visitors across both time and the environmental space (Hardy & Aryal, 2019).

Finally, the future use of automation to enhance the *in situ* visitor experience is gaining traction in academic research. The development of virtual tour guides within attractions that intuitively mimic human interactions provides new levels of autonomy for visitors while minimising labour costs (Bickmore *et al.*, 2013). Extending this concept into the future lies in the advancement of artificial intelligence capabilities where virtual (or indeed robotic) guides can tailor content to specific audience types (such as children, families, users with sensory or mobility impairments). Furthermore, a critical future trajectory is the integration of authentic storytelling beyond visitor instruction. While current prototypes can communicate facts and standardised stories, future innovations may allow for further personalisation depending on the user characteristics. Similarly, existing mobile-enabled guides based on global positioning technology provide targeted information to visitors often in outdoor attraction settings (Chu *et al.*, 2012). However, these still rely on smartphone devices to deliver content. Future integration of global positioning and beacon technology may allow for the enhancement of these tools into wearable platforms (such as smart glasses, smart watches or smart lenses) or implanted technology. The future of virtual and augmented experiences poses significant opportunities for attractions where their longevity may be in doubt. Particularly in the natural and built heritage landscape, current assets and resources are finite and future travellers may utilise virtual re-creation more to access sites threatened with degradation or decay (Itani & Hollebeek, 2021).

5.3.4 Food and beverage sector

Automation within the food and beverage sector has already experienced significant growth. The presence of digital menus, online ordering and reservation systems is commonplace in a variety of hospitality settings. The future trajectories for automation in this sector are dominated by two forces: self-service technology and robotic service. As suggested by Wei *et al.* (2017), the demand for self-service and enhanced customisation is a major driving force for the food and beverage sector. While self-service is not new (fast-food chains have been experimenting with this for several years), the future for this concept is diverse. For example, the growth of chatbot applications where guests can rapidly customise and place specific requests in real time is already occurring in hotels; however, this is likely to expand into food and beverage settings in the future (Um *et al.* 2020). Additional self-service concepts such as autonomous ordering without human mediation exist, although these are limited due to technological device functionality. For instance, Gummerus *et al.* (2019) questions how self-service devices may become further integrated into customers' daily lives and physical presence. Future developments may revolve around the removal of 'self-service devices', where ordering, transaction and collection are mediated by emerging smart devices or spaces. As discussed by Buhalis (2020), the future development of ambient intelligence through a combination of disruptive technologies including sensor/beacon networks, cryptocurrency, RFID, interconnected networks and the internet of things provides opportunities for seamless control. Voice control, smart surfaces and holographic projection will provide opportunities for consumers to purchase, customise and adapt products within their own ambient space. Similar to the MagicBand concept in attractions, future developments in this sector may explore the use of biometric or wearable technology to facilitate self-service or the use of particular sensory triggers such as eye tracking within technology-mediated food and beverage servicescapes.

The expansion of robotics within the food and beverage services has received significant interest in recent years. As noted by Jang and Lee (2020), serving robots perceived as likeable, intelligent and safe can be highly valued in the minds of food and beverage consumers. However, in the future, the consumer 'wow factor' may reconfigure. While currently service robots hold a novelty value for consumers, with mass adoption and deployment across the sector, the future 'wow factor' may emerge through social presence. As discussed by Wirtz *et al.* (2018), as customers gradually accept robotics into their daily lives, value is likely to transfer back to the human, relational interaction and the social contact that is embodied by food and beverage personnel. While at various levels of development from mechanical (such as the Bionic Bar onboard Royal Caribbean Cruise Lines) to humanoid, the future trajectory for automation is likely to move towards value-adding robotics as a point of differentiation. In the case of robotic chefs, Fusté-Forné (2021) argued that these would largely be resigned to the fast-food industry where standardised practices and rapid efficiency are paramount to the business model. However, future developments may include the integration of robotics into the fine dining or 'slow food' environment to supplement the valued human contact afforded by talented hospitality personnel.

This may lead to hybrid experiences that combine the efficiency benefits of robotics with the spontaneity of human professionals. As discussed by Wirtz *et al.* (2018), the future human–robot divide within services will be dictated by task complexity. Robotics are more likely to assimilate tasks with high cognitive or analytical requirements that fulfil functional needs (e.g. ticketing services, transaction, information provision). Conversely, high emotional or social tasks that require authentic interaction, engagement or relational needs are likely to remain delivered by human service professionals in the future.

Further future trajectories for food and beverage robotics include the advancement of 'human-like' behaviour. Anthropomorphic robotics featuring human likeness including language, style, voice and appearance have been seen (even in existing settings) as more reliable and relatable to the public (Lu *et al.*, 2021). As such, future developments in this area will likely involve the enhanced design of robotic servers to further put customers at ease and to blur the visible divide between human and robot personnel. Similarly, advances in motion and detection are anticipated to be significant growth areas. Robotic servers of the future are likely to exhibit greater ranges of movement to flow more safely and seamlessly throughout the space, in addition to advanced detection systems to anticipate and minimise collision with guests (Wan *et al.*, 2020).

5.4 Future Tourism: Challenges and Opportunities in an Age of Automation

As suggested by Samala *et al.* (2020), the future growth of automated technologies will lead to significant changes for tourism businesses, their processes and the overall industry landscape. Change will also be necessary for employees working within the sector and, indeed, consumers who engage with those services. Returning to Figure 5.1, the following discussion explores the challenges and opportunities for future tourism in an automated age at the business, employee and consumer levels.

5.4.1 Business level

At the business and managerial level, tourism and hospitality practitioners will undoubtedly adapt their strategic approach with the growth of automated technology. As noted earlier in the chapter, the uncertain availability of human labour, skills and workforce in tourism coupled with the rapid advancement in technological capabilities indicates a substantial shift in management approach in the future. Xu *et al.* (2020) argue that tourism businesses of the future will require a change in leadership ethos to accommodate and embrace automated technology and robotics. A critical challenge for leaders will be how to leverage automation alongside human resources. While the concept of purely robot-based organisations may be pursued in the long term, the journey to that will involve smaller-scale incremental adoptions where technology is further used to complement the soft skills of the workforce (Xu et al., 2020). In the case of robotics, Ivanov et al. (2020a) suggest that the key impact for managers will be the perceived streamlining of operational processes and the increased capacity that robotics offers. Similarly, alternative automated technologies (such as those explored above) may be intriguing for businesses; however, they equally need to provide adequate cost savings and efficiencies to be lucrative to tourism and hospitality managers (Rozila & Scott, 2019). This is particularly relevant for future back-of-house operations where routine operational tasks will increasingly be transferred to automated systems (Mahey, 2020). Process-based activities such as bookkeeping, filing, data storage, human resource reporting and financial accounting will increasingly move to automated entities to allow human employees greater time to focus on relational management tasks in the service delivery process. From a future perspective, smaller-scale integration offers the opportunity for these benefits to be observed by businesses, which will undoubtedly lead to greater willingness to adapt. A key capability to ease future transition is to further develop the technology producer–service provider–customer relationship (Ivkov et al., 2020) to integrate various stakeholders in the design and development of automated technologies for context-specific services.

From an operational perspective, the concern over value for money from tourism and hospitality businesses may inhibit growth in the adoption of automated technology. However, the cost of automated technology will likely fall in the future with more start-up businesses advancing the quality and functionality of technologies while also operating more competitively in the market (leading to a reduction in costs). Huang and Rust (2018) identified that with an increase in technological capabilities, automation gradually becomes a more effective substitute for human labour and, as such, productivity increases exponentially. If such findings can be replicated in tourism and hospitality, it is anticipated that tourism leaders will explore automation with greater impetus into the future. However, this would reconfigure the budget and luxury markets of the future. With the increased savings of fully automated providers, these spaces may become dominated by the budget traveller valuing speed, efficiency and minimal variation. In contrast, luxury travellers of the future may purchase experiences with greater human contact. The prices of these will likely increase and change perceptions of value from amenities to human contact and socialisation within the tourism experience.

5.4.2 Employee level

The future for employees in tourism and hospitality is an area of significant debate. Melián-González and Bulchand-Gidumal's (2020) trend analysis suggests that the move to workplace automation in hospitality has already begun and has increased over the past 10 years. In casting forward into the future, it is likely that the divide between automation and employees will narrow as a result of more roles and duties being incorporated into automated platforms. As suggested by Ivanov (2020), the automated revolution will undoubtedly cost jobs (through the removal of certain roles), change jobs (to inform and integrate with automated platforms) and create new jobs (such as maintenance, designers and user-experience specialists within tourism/hospitality firms). As discussed by Ivanov *et al.* (2020b), there is reason to fear automation from the employee perspective. Not only in the case of workers being replaced by technology but also in finding new roles within the same or other industries. Such a fear is heightened among the older workforce where adapting to a new automation job market may be met with apprehension. For employees of the future, this will not only create a paradigm shift in tourism employability but also offer significant potential for growing skills, innovation and creativity.

The growth of technological capabilities will also reconfigure the traditional tourism/hospitality skills mix. As argued by Autor (2015), progressive automation in the future will likely influence an unbundling of skills and duties. For tourism and hospitality employees, routine and/or mundane tasks will most likely be devolved to automated entities where tasks requiring greater personal contact or rapid adaptability may be reserved for human employees. Therefore, a critical future challenge for employees will not only be about how they work individually and in teams, but also about how they work with non-human or automated entities in the workplace. As such, new personal skills will be required and tourism/hospitality educators must consider which 'future-proof' skills may be required in a hybridised human/robotic workplace (El Hajal & Rowson, 2020). Similarly, the integration of automated technology into tourism and hospitality will likely raise new and unique training requirements for employees both in how they use the platforms and how they can best integrate them into the service delivery process (Bowen & Morosan, 2018). This poses interesting debates about the future of recruitment, training and development within the tourism landscape to ensure that the human workforce is equipped to coexist with robotic personnel (Wirtz *et al.*, 2018).

5.4.3 Consumer level

For tourism and hospitality consumers, future automations will radically reconfigure the service experience. Kabadayi *et al.* (2019) argued that smart technology has the potential to offer convenience, information and social and emotional value for consumers. As automation becomes more commonplace within a range of tourism and hospitality settings, it is conceivable that it will also become more firmly embedded in the value perception of consumers (Naumov, 2019) and arguably become an expectation for the service offering. Future consumers may therefore reconfigure their perception of technology as a mediating force – to another 'actor' within the experience capable of offering co-creative opportunities (Sun Tung & Au, 2018).

A critical challenge from the consumer perspective will be in relation to trust. As researched by Park (2020), consumers determine their level of trust in a platform based on three components: performance (the ability and capability of the entity); process (the means by which they operate); and purpose (why the platform was developed and for what benefit). Currently, with automation and robotics still on the periphery of mainstream tourism and hospitality experiences, it may be difficult for consumers to evaluate platforms and thereby attribute trust to them. As highlighted by Wirtz *et al.* (2018), some of the trust concerns associated with automation and robotics stem from worries over safety, privacy and data security. This is particularly relevant in the case of mistakes or malfunctions where trust in using the technology may be reduced. As such, consumers of the future will need to assimilate new roles, skills and competencies not only to engage with automation but also to be supported in how to use such innovations safely (Mingotto *et al.*, 2020). In the future, with more of these technologies becoming visible, it will be critical for businesses to showcase and communicate the benefits, performance and purpose of intelligent machines to garner positive consumer sentiment (Tussyadiah *et al.*, 2020). This can be complemented by wider societal change focusing on greater governance over automated technologies to ensure ethical, security and data protection standards (Wright & Schultz, 2018). With new layers of protection and enforcement, consumers of the future are more likely to embrace the power of automation and build an appropriate level of trust in tourism and hospitality systems.

5.5 Conclusion

This chapter sought to consider the future trajectories of service automation in a range of tourism and hospitality settings. Through consideration of developing technologies including autonomous platforms, data-driven applications and robotics, a variety of new opportunities were identified. Transport, accommodation, attractions and food and beverage providers have significant potential to adopt and integrate automated technology into their service provision; however, challenges may occur. The advancement of technology will inevitably bring revolutionary changes to the tourism experience and require a significant reconfiguration of existing tourism and hospitality business practice. As technology in future tourism may move beyond the role of mediator to that of an equal actor in the tourism/hospitality experience, a key theme for the future of tourism in a robot-based economy lies in interconnectivity. The growth in the internet of things and Web 3.0 will streamline the tourism experience with currently disparate systems and processes amalgamating throughout the purchase journey. Experiential spaces will need to be reconfigured not only to accommodate the hardware of the future but also to provide the customer-facing platforms with which future tourists will engage. Smart surfaces, projection and holographic technology will feature more heavily, and reliance on tangible devices (such as smartphones and tablets) will begin to move into ambient technology. The future of the tourism and hospitality landscape will grant even more autonomy to customers where elements of the experience can be synchronously adapted and customised through voice recognition and biometric scanning. Beyond this, the prospect of technology implantation to capture and share data, communicate and interact with service organisations poses significant questions about human–technology hybridisation. Undoubtedly, this will change how customers engage with services. Current fears surrounding safety, data security, integrity and trust may remain, but with the wider proliferation of automated technology and enhanced customer skills, these fears are likely to reduce. Adapting to the future of automated technology may cause challenges for tourism and hospitality businesses. The balance between human personnel and automated systems must be managed carefully to capture the potential efficiencies while retaining the social interaction and relationships that future customers may continue to value. While further research is needed to explore and evaluate this progression, this theoretical work provides a foundation for the applications, challenges and opportunities afforded by automated technology for future tourism activity.

References

Autor, D. (2015) Why are there still so many jobs? The history and future of workplace automation. *Journal of Economic Perspectives* 29 (3), 3–30. https://doi.org/10.1257/jep.29.3.3

Balasubramanian, K. and Ragavan, N. (2019) What are the key challenges faced by the Malaysian hospitality and tourism industry in the context of industrial revolution 4.0? *Worldwide Hospitality and Tourism Themes* 11 (2), 194–203. https://doi.org/10.1108/WHATT-11-2018-0079

Baum, T. (2019) Does the hospitality industry need or deserve talent? *International Journal of Contemporary Hospitality Management* 31 (10), 3823–3837. https://doi.org/10.1108/IJCHM-10-2018-0805

Bickmore, T., Vardoulakis, L. and Schulman, D. (2013) Tinker: A relational agent museum guide. *Autonomous Agents and Multi-Agent Systems* 27 (2), 254–276. https://doi.org/10.1007/s10458-012-9216-7

Bolton, C., Machová, V., Kovacova, M. and Valaskova, K. (2018) The power of human–machine collaboration: Artificial intelligence, business automation, and the smart economy. *Economics, Management and Financial Markets* 13 (4), 51–56. https://doi.org/10.22381/EMFM13420184

Borkowski, S., Sandrick, C., Wagila, K., Goller, C., Ye, C. and Zhao, L. (2016) MagicBands in the magic kingdom: Customer-centric information technology implementation at Disney. *Journal of the International Academy for Case Studies* 22 (4), 143–151.

Bowen, J. and Morosan, C. (2018) Beware hospitality industry: The robots are coming. *Worldwide Hospitality and Tourism Themes* 10 (6), 726–733. https://doi.org/10.1108/WHATT-07-2018-0045

Buhalis, D. (2020) Technology in tourism-from information communication technologies to eTourism and smart tourism towards ambient intelligence tourism: A perspective article. *Tourism Review* 75 (1), 267–272. https://doi.org/10.1108/TR-06-2019-0258

Buhalis, D. and Sinarta, Y. (2019) Real-time co-creation and nowness service: Lessons from tourism and hospitality. *Journal of Travel & Tourism Marketing* 36 (5), 563–582. https://doi.org/10.1080/10548408.2019.1592059

Buhalis, D., Harwood, T., Bogicevic, V., Viglia, G., Beldona, S. and Hofacker, C. (2019) Technological disruptions in services: Lessons from tourism and hospitality. *Journal of Service Management* 30 (4), 484–506. https://doi.org/10.1108/JOSM-12-2018-0398

Chu, T., Lin, M. and Chang, C. (2012) mGuiding (Mobile Guiding): Using a mobile GIS app for guiding. *Scandinavian Journal of Hospitality and Tourism* 12 (3), 269–283. https://doi.org/10.1080/15022250.2012.724921

Cohen, S. and Hopkins, D. (2019) Autonomous vehicles and the future of urban tourism. *Annals of Tourism Research* 74, 33–42. https://doi.org/10.1016/j.annals.2018.10.009

Economic Insight (2019) Hospitality and tourism workforce landscape. Economic Insight Limited Report. https://www.economic-insight.com/wp-content/uploads/2019/10/DCMS-final-annex-13-06-19-STC.pdf (accessed 23 January 2022).

El Hajal, G. and Rowson, B. (2020) The future of hospitality jobs. *Research in Hospitality Management* 10 (1), 55–61. https://doi.org/10.1080/22243534.2020.1790210

Fusté-Forné, F. (2021) Robot chefs in gastronomy tourism: What's on the menu? *Tourism Management Perspectives* 37, 100774. https://doi.org/10.1016/j.tmp.2020.100774

Gretzel, U., Sigala, M., Xiang, Z. and Koo, C. (2015) Smart tourism: Foundations and developments. *Electronic Markets* 25 (3), 179–188. https://doi.org/10.1007/s12525-015-0196-8

Gretzel, U., Fuchs, M., Baggio, R., Hoepken, W., Law, R., Neidhardt, J., Pesonen, J., Zanker, M. and Xiang, Z. (2020) E-Tourism beyond COVID-19: A call for transformative research. *Information Technology & Tourism* 22 (2), 187–203. https://doi.org/10.1007/s40558-020-00181-3

Gummerus, J., Lipkin, M., Dube, A. and Heinonen, K. (2019) Technology in use: Characterizing customer self-service devices (SSDS). *The Journal of Services Marketing* 33 (1), 44–56. https://doi.org/10.1108/JSM-10-2018-0292

Guo, Y., Liu, H. and Chai, Y. (2014) The embedding convergence of smart cities and tourism internet of things in China: An advance perspective. *Advances in Hospitality and Tourism Research* 2 (1), 54–69.

Hansen, C., Daim, T., Ernst, H. and Herstatt, C. (2016) The future of rail automation: A scenario-based technology roadmap for the rail automation market. *Technological Forecasting & Social Change* 110, 196–212. https://doi.org/10.1016/j.techfore.2015.12.017

Hardy, A. and Aryal, J. (2019) Using innovations to understand tourist mobility in national parks. *Journal of Sustainable Tourism* 28 (2), 263–283. https://doi.org/10.1080/09669582.2019.1670186

Huang, M. and Rust, R.T. (2018) Artificial intelligence in service. *Journal of Service Research* 21 (2), 155–172. https://doi.org/10.1177/1094670517752459

Itani, O. and Hollebeek, L.D. (2021) Light at the end of the tunnel: Visitors' virtual reality (versus in-person) attraction site tour-related behavioral intentions during and post-COVID-19. *Tourism Management* 84, 104290. https://doi.org/10.1016/j.tourman.2021.104290

Ivkov, M., Blešić, I., Dudić, B., Bartáková, G.P. and Dudić, Z. (2020) Are future professionals willing to implement service robots?: Attitudes of hospitality and tourism students towards service robotization. *Electronics* 9 (9), 1442. https://doi.org/10.3390/electronics9091442

Ivanov, S. (2020) The impact of automation on tourism and hospitality jobs. *Information Technology & Tourism* 22 (2), 205–215. https://doi.org/10.1007/s40558-020-00175-1

Ivanov, S. and Webster, C. (2017) Designing robot-friendly hospitality facilities. *Proceedings of the Scientific Conference 'Tourism. Innovations. Strategies'*, 13–14 October, Bourgas, Bulgaria, pp. 74–81, https://ssrn.com/abstract=3053206, SSRN Publishing.

Ivanov, S. and Webster, C. (2019) Conceptual framework of the use of robots, artificial intelligence and service automation in travel, tourism, and hospitality companies. In S. Ivanov and C. Webster (eds) *Robots, Artificial Intelligence, and Service Automation in Travel, Tourism and Hospitality* (pp. 7–37). Emerald Publishing. https://doi.org/10.1108/978-1-78756-687-320191007

Ivanov, S., Seyitoğlu, F. and Markova, M. (2020a) Hotel managers' perceptions towards the use of robots: A mixed-methods approach. *Information Technology & Tourism* 22, 505–535. https://doi.org/10.1007/s40558-020-00187-x

Ivanov, S., Kuyumdzhiev, M. and Webster, C. (2020b) Automation fears: Drivers and solutions. *Technology in Society* 63, 101431. https://doi.org/10.1016/j.techsoc.2020.101431

Jang, H. and Lee, S. (2020) Serving robots: Management and applications for restaurant business sustainability. *Sustainability* 12 (10), 3998–4013. https://doi.org/10.3390/su12103998

Kabadayi, S., Ali, F., Choi, H., Joosten, H. and Lu, C. (2019) Smart service experience in hospitality and tourism services: A conceptualization and future research agenda. *Journal of Service Management* 30 (3), 326–348. https://doi.org/10.1108/JOSM-11-2018-0377

Keuchel, S. (2020) Digitalisation and automation of transport: A lifeworld perspective of travellers. *Transportation Research Interdisciplinary Perspectives* 7, 100195. https://doi.org/10.1016/j.trip.2020.100195

Kim, H. and Kim, B. (2018) A qualitative approach to automated motels: A rising issue in South Korea. *International Journal of Contemporary Hospitality Management* 30 (7), 2622–2636. https://doi.org/10.1108/IJCHM-03-2017-0127

Kim, J. and Han, H. (2020) Hotel of the future: Exploring the attributes of a smart hotel adopting a mixed-methods approach. *Journal of Travel & Tourism Marketing* 37 (7), 804–822. https://doi.org/10.1080/10548408.2020.1835788

Kuo, C., Chen, L. and Tseng, C. (2017) Investigating an innovative service with hospitality robots. *International Journal of Contemporary Hospitality Management* 29 (5), 1305–1321. https://doi.org/10.1108/IJCHM-08-2015-0414

Loureiro, A. (2017) How technology is successfully transforming travel to better serve the ever-connected digital consumer. *Worldwide Hospitality and Tourism Themes* 9 (6), 675–678. https://doi.org/10.1108/WHATT-09-2017-0058

Lu, L., Zhang, P. and Zhang, T. (2021) Leveraging 'human-likeness' of robotic service at restaurants. *International Journal of Hospitality Management* 94, 102823. https://doi.org/10.1016/j.ijhm.2020.102823

Mahey, H. (2020) *Robotic Process Automation with Automation Anywhere: Techniques to Fuel Business Productivity and Intelligent Automation Using RPA*. Packt Publishing.

Manthiou, A., Klaus, P., Kuppelwieser, V.G. and Reeves, W. (2020) Man vs machine: Examining the three themes of service robotics in tourism and hospitality. *Electronic Markets* 31 (3), 511–527. https://doi.org/10.1007/s12525-020-00434-3

Melián-González, S. and Bulchand-Gidumal, J. (2020) Employment in tourism: The jaws of the snake in the hotel industry. *Tourism Management* 80, 104123. https://doi.org/10.1016/j.tourman.2020.104123

Mingotto, E., Montaguti, F. and Tamma, M. (2020) Challenges in re-designing operations and jobs to embody AI and robotics in services. Findings from a case in the hospitality industry. *Electronic Markets* 31 (3), 493–510. https://doi.org/10.1007/s12525-020-00439-y

Mygind, L. and Bentsen, P. (2017) Reviewing automated sensor-based visitor tracking studies: Beyond traditional observational methods? *Visitor Studies* 20 (2), 202–217. https://doi.org/10.1080/10645578.2017.1404351

Naumov, N. (2019) The impact of robots, artificial intelligence, and service automation on service quality and service experience in hospitality. In S. Ivanov and C. Webster (eds) *Robots, Artificial Intelligence, and Service Automation in Travel, Tourism and Hospitality* (pp. 123–133). Emerald Publishing. https://doi.org/10.1108/978-1-78756-687-320191007

Nikitas, M., Michalakopoulou, K., Njoya, E. and Karampatzakis, D. (2020) Artificial intelligence, transport and the smart city: Definitions and dimensions of a new mobility era. *Sustainability* 12 (7), 2789. https://doi.org/10.3390/su12072789

Panagiotopoulos, I. and Dimitrakopoulos, G. (2018) An empirical investigation on consumers' intentions towards autonomous driving. *Transportation Research. Part C, Emerging Technologies* 95, 773–784. https://doi.org/10.1016/j.trc.2018.08.013

Papadimitriou, E., Schneider, C., Aguinaga T., Juan, D., Wouter, L.V.M. and Ten Broeke, A. (2020) Transport safety and human factors in the era of automation: What can transport modes learn from each other? *Accident Analysis and Prevention* 144, 105656. https://doi.org/10.1016/j.aap.2020.105656

Park, S. (2020) Multifaceted trust in tourism service robots. *Annals of Tourism Research* 81, 102888. https://doi.org/10.1016/j.annals.2020

Prideaux, B. and Yin, P. (2019) The disruptive potential of autonomous vehicles (AVs) on future low-carbon tourism mobility. *Asia Pacific Journal of Tourism Research* 24 (5), 459–467. https://doi.org/10.1080/10941665.2019.1588138

Reid, J. (2018) St Pancras gets UK travel industry's first humanoid robot. *Business Traveller* [online]. https://www.businesstraveller.com/rail-travel/2018/10/24/st-pancras-gets-uk-travel-industrys-first-humanoid-robot/ (accessed 12 January 2021).

Ristova, C. and Dimitrov, N. (2019) Digitalization in the hospitality industry: Trends that might shape the next stay of guests. *International Journal of Information, Business and Management* 11 (3), 144–154.

Ritzer, G. (2015) Hospitality and prosumption. *Research in Hospitality Management* 5 (1), 9–17. https://doi.org/10.1080/22243534.2015.11828323

Rozila, A. and Noel, S. (2019) Technology innovations towards reducing hospitality human resource costs in Langkawi, Malaysia. *Tourism Review* 74 (3), 547–562. https://doi.org/10.1108/TR-03-2018-0038

Samala, N., Katkam, B.S., Bellamkonda, R.S. and Rodriguez, R.V. (2020) Impact of AI and robotics in the tourism sector: A critical insight. *Journal of Tourism Futures* 8 (1), 73–87. https://doi.org/10.1108/JTF-07-2019-0065

Seal, P. (2019) guest retention through automation: An analysis of emerging trends in hotels in Indian sub-continent. In D. Batabyal and D.K. Das (eds) *Global Trends, Practices, and Challenges in Contemporary Tourism and Hospitality Management* (pp. 58–69). IGI Global. https://doi.org/10.4018/978-1-5225-8494-0.ch003

Sun Tung, V.W. and Au, N. (2018) Exploring customer experiences with robotics in hospitality. *International Journal of Contemporary Hospitality Management* 30 (7), 2680–2697. https://doi.org/10.1108/IJCHM-06-2017-0322

Tung, V. and Law, R. (2017) The potential for tourism and hospitality experience research in human–robot interactions. *International Journal of Contemporary Hospitality Management* 29 (10), 2498–2513. https://doi.org/10.1108/IJCHM-09-2016-0520

Tuominen, P. and Ascenção, M. (2016) The hotel of tomorrow: A service design approach. *Journal of Vacation Marketing* 22 (3), 279–292. https://doi.org/10.1177/1356766716637102

Tussyadiah, I. (2020) A review of research into automation in tourism: Launching the *Annals of Tourism Research* curated collection on artificial intelligence and robotics in tourism. *Annals of Tourism Research* 81,102883, 1–13. https://doi.org/10.1016/j.annals.2020.102883

Tussyadiah, I., Zach, F. and Wang, J. (2020) Do travelers trust intelligent service robots? *Annals of Tourism Research* 81, 102886. https://doi.org/10.1016/j.annals.2020.102886

Um, T., Kim, T. and Chung, N. (2020) How does an intelligence chatbot affect customers compared with self-service technology for sustainable services? *Sustainability* 12 (12), 5119. https://doi.org/10.3390/su12125119

Urquhart, E. (2019) Technological mediation in the future of experiential tourism. *Journal of Tourism Futures* 5 (2), 120–126. https://doi.org/10.1108/JTF-04-2019-0033

Wan, A., De Soong, Y., Foo, E., Wong, W. and Lau, W. (2020) Waiter robots conveying drinks. *Technologies* 8 (3), 44–59. https://doi.org/10.3390/technologies8030044

Wang, X., Li, X., Zhen, F. and Zhang, J. (2016) How smart is your tourist attraction?: Measuring tourist preferences of smart tourism attractions via a FCEM-AHP and IPA approach. *Tourism Management* 54, 309–320. https://doi.org/10.1016/j.tourman.2015.12.003

Webster, C. (2021) Demography as a driver of robonomics. *ROBONOMICS: The Journal of the Automated Economy* 1, 12. https://journal.robonomics.science/index.php/rj/article/view/12

Webster, C. and Ivanov, S. (2020) Demographic change as a driver for tourism automation. *Journal of Tourism Futures* 6 (3), 263–270. https://doi.org/10.1108/JTF-10-2019-0109

Wei, W., Torres, E. and Hua, N. (2017) The power of self-service technologies in creating transcendent service experiences: The paradox of extrinsic attributes. *International Journal of Contemporary Hospitality Management* 29 (6), 1599–1618. https://doi.org/10.1108/IJCHM-01-2016-0029

Wirtz, J., Patterson, P.G., Kunz, W.H., Gruber, T., Lu, V.N., Paluch, S. and Martins, A. (2018) Brave new world: Service robots in the frontline. *Journal of Service Management* 29 (5), 907–931. https://doi.org/10.1108/JOSM-04-2018-0119

Wright, S. and Schultz, A. (2018) The rising tide of artificial intelligence and business automation: Developing an ethical framework. *Business Horizons* 61 (6), 823–832. https://doi.org/10.1016/j.bushor.2018.07.001

Xu, S., Stienmetz, J. and Ashton, M. (2020) How will service robots redefine leadership in hotel management? A Delphi approach. *International Journal of Contemporary Hospitality Management* 32 (6), 2217–2237. https://doi.org/10.1108/IJCHM-05-2019-0505

Zhang, W., Jenelius, E. and Badia, H. (2019) Efficiency of semi-autonomous and fully autonomous bus services in trunk-and-branches networks. *Journal of Advanced Transportation* 019, 1–17. https://doi.org/10.1155/2019/7648735

6 The Automated Tourism and Hospitality Company of the Future

Stanislav Ivanov and May Kristin Vespestad

6.1 Introduction

In the robonomic economic system, many tourism and hospitality companies will be completely automated, and some may have no human employees. This means that we must rethink the way tourism attractions are created, the way tourism experiences are designed and the practices of tourism and hospitality companies. We need to consider how to cater for tourists in ways that prove viable both with and by (ro)bots, as well as with human employees who might have a changed role in future tourism and hospitality. The automated experiencescape of the future can be envisaged as that of changed roles and practices, which if designed with care, could also contribute to more sustainable solutions within the tourism and hospitality industry.

In the future, tourism and hospitality companies may be owned and run by artificial intelligences (AIs) and may have no human employees. The AI is not owned by someone, unlike today when technology is an asset owned by someone. In the future, AI could be legally treated as animals, i.e. some have owners, others do not. In future tourism and hospitality, experiences may be (co-)created by humans and/or AI. This will likely lead to a 'rational–emotional clash' when the rational decision-making of AI meets the emotional reactions of human tourists. To counteract such conflicts, front-stage AI must have emotional intelligence. The automation of future tourism will indeed influence the co-creation of experiential value for all stakeholders.

In the event of human-out-of-the-loop decision-making approaches, where no human is involved, decisions are taken and implemented by AI. Tourism and hospitality management decisions relating to operational practices, marketing and financial aspects of the automated tourism and hospitality companies will be different from current norms. For example, AI can communicate with other AIs in the supply chain. At the same time, the need for human involvement will be evident, particularly in tourism companies' operation in an unpredictable context (e.g. outdoors experience-based companies) or highly interactive emotional settings such as complex problem-solving relying on highly specialised human competence (e.g. gourmet chefs).

This chapter focuses on the characteristics of the automated tourism and hospitality company of the future. It elaborates on the automation of tasks and processes in tourism and hospitality and the partial and full automation of the industry, and sheds light on the ways business processes are organised in automated tourism and hospitality companies in the future.

6.2 Automating Tasks and Processes in Tourism and Hospitality

From an operations management perspective, every tourism and hospitality organisation is a bundle of processes (Slack *et al.*, 2022), e.g. information provision, booking a room/cabin/seat/table, cooking, serving, cleaning, housekeeping, pricing, document handling, processing payments from guests, paying suppliers and inventory management. The processes are usually divided into back-of-house (or back-stage) and front-of-house (or front-stage) processes based on customer participation in them. This division into back-of-house (without customer participation and co-creation) and front-of-house (with customer participation and co-creation) processes determines the degree of control that managers have over the respective processes and their potential automatability. Managers and employees have full control over the back-of-house processes; they are usually repetitive, have a predictable duration and outcomes, can often be planned in detail, and therefore are automatable provided the availability of appropriate technology. The inclusion of the customer in front-of-house processes increases the uncertainty of the process procedures, duration and outcome, and therefore makes the process generally more difficult to automate because customers may resist the use of automation (Webster & Ivanov, 2021) and may not follow service delivery procedures (e.g. they may show aggression towards a waiter robot). It becomes even more challenging to streamline the co-creation processes of memorable experiences.

The road to full automation in tourism goes through the automation of processes in tourism and hospitality companies. Each process is a bundle of tasks. Hence, the automation of processes happens in practice through the automation of tasks that constitute these processes (Ivanov, 2020). The automation of tasks in tourism and hospitality depends on factors such as the characteristics of the technological solutions available to companies, cost considerations, target customer preferences and the automatability of tasks. Advances in robotics, AI and other automation technologies increase the capabilities of these technologies and make the automation of more and more tasks technologically feasible while the decreasing technological costs make task automation affordable.

Customers' preferences for human-delivered authentic services may be a hindrance to full automation (Seyitoğlu, 2021). However, even if tourism and hospitality companies can afford to implement the latest technologies and customers are eager to have automated experiences, the automatability of tasks in tourism and hospitality rises as a barrier to full automation. Task automatability refers to 'how easy it is for a task to be performed by technology rather than by a human employee' (Ivanov, 2023). A task's automatability depends on the task's characteristics: nature, complexity, frequency and standardisation.

Based on their *nature*, tasks can be grouped into physical (e.g. moving items, cleaning, cooking) and cognitive (e.g. providing information, recording a booking into the property management system of a hotel, issuing documents, generating draft contracts) tasks. Both groups of tasks are automated with the help of different technologies. Performing physical tasks requires mobility and the physical handling of objects (humans, luggage, food, etc.). In the tourism and hospitality context, they are automated through industrial, service or social robots, e.g. for cooking food, for serving food, for room service delivery, for cleaning floors and for disinfecting premises (Tuomi *et al.*, 2021). Cognitive tasks can be automated through software/intelligent automation (Bornet *et al.*, 2021) which is not necessarily embedded in a robotic device. Physical tasks are more complex than cognitive tasks because the robot that performs them needs to gather, analyse and respond quickly to real-time data from its physical surroundings. Therefore, by their nature, the physical tasks in tourism and hospitality incorporate cognitive elements too. At the same time, cognitive tasks may not need the physical handling of objects and mobility (e.g. extracting information from an email message and inputting in the property management system of a hotel). However, some cognitive tasks such as communicating with customers require the physical movement of the head, hands and torso of the social robot and maintaining eye contact with the guest to improve communication between the robot and the guest, since not all communication is verbal. Finally, many tasks in tourism and hospitality companies (e.g. handling customer complaints, dealing with emergency situations) require high emotional intelligence from the service provider. The current limited emotional intelligence capabilities of automation technologies make such tasks very difficult to automate; however, in the robonomic economy of the future, the emotional skills of chatbots and robots will be sufficient to automate them as well.

Task *complexity* is 'the aggregation of any intrinsic task characteristic that influences the performance of a task' (Liu & Li, 2012: 559). These include:

- Goal: What the task-doer wants to achieve with the task (e.g. moving the ordered dishes from the kitchen to the guests' table).

- Input: Quantity, diversity, predictability, variability, duration, regularity and other characteristics of a task's inputs (e.g. fluctuation in demand, serving hotel guests with conflicting service requirements).
- Procedure: The way a task is performed, the number of required steps (e.g. scanning a guest's ID card, extracting data from it and inputting it in the property management system or typing the data on a computer's keyboard).
- Output: The result of the task in terms of quantity and quality (e.g. customer's data input correctly).
- Time: Duration of the task and time pressure/urgency for its completion.
- Presentation: The way the output is to be presented (e.g. formal or casual table setting).

Based on their complexity, tasks can be divided into groups between two extremes – simple and complex tasks. Simple tasks have clear goals, predicable input, take only a few steps, have a clear output definition that does not require much effort and may be short in duration. Complex tasks require high levels of critical thinking and creativity by the task-doer because they involve many steps, have variable and diverse inputs, not well-defined procedures and outputs and may have unpredictable duration. Moreover, a complex task may require interaction between several task-doers (e.g. employees) or with customers (i.e. co-creation of experiences). Simple tasks (e.g. inputting customer data in a database) are generally easier to automate compared to complex tasks (e.g. restaurant kitchen layout design); however, advances in generative AI allow for the automation of many complex tasks in tourism and hospitality as well (Carvalho & Ivanov, 2024; Iskender, 2023).

Task *frequency* is a characteristic that divides tasks based on how often they need to be implemented in a tourism/hospitality company. Repetitive tasks are worth automating because they create economies of scale due to the automation of numerous tasks, often across job positions. For example, for a hotel it is worth investing in chatbot technology to automate bookings and answer customer requests. Tasks that are rarely performed in a company are not economically feasible to automate.

Task *standardisation/algorithmisation* refers to the variability of a task and the possibility to develop a strict procedure for its implementation. Tasks with a well-defined and consistent procedure are automatable because there is little variation in the tasks' inputs and output (e.g. preparing fries in a fast-food restaurant). Tasks with high variability (e.g. preparing dishes in a fine dining à la carte restaurant) are more difficult to automate. Nevertheless, automation technologies already allow for a high degree of customisation and personalisation (e.g. dynamic packaging) which will be the norm in the robonomic economy of the future, although customers may not always wish to pay more for customised and personalised tourism and hospitality products (Ivanova *et al.*, 2021).

Simple, standardised and repetitive tasks are highly automatable, while complex, rare and diverse tasks are more difficult to automate. However, advances in automation technologies widen the scope of automatable tasks, paving the way to full automation in tourism and hospitality. The latter will require a reorganisation of processes within tourism and hospitality companies – some will disappear because they are unnecessary, others will incorporate new or fewer tasks, and a third will be newly created processes (Ivanov, 2020). The human service providers will gradually step aside while automation technologies will take their place on the scene, although human service providers will still be employed by some tourism and hospitality companies.

6.3 Partial vs Full Automation in Tourism and Hospitality

For tourism and hospitality companies, full and partial automation have their own advantages and disadvantages. Most importantly, partial automation compensates for the weaknesses of automation technologies (e.g. lack of flexibility), using the strengths of human employees (e.g. creativity, critical thinking, emotional intelligence, resourcefulness) and vice versa. This means that customers have the support of human employees when needed (e.g. if a machine malfunctions or they do not know how to use it) or if they want to be served by humans rather than robots (Seyitoğlu & Ivanov, 2020). However, the ironies of automation work against partial automation (see Chapter 2). According to Bainbridge (1983), automation may create more problems than eliminate them due to errors in the system design and the arbitrary collection of tasks human operators are left with that system designers could not automate. For example, engineers might automate the cooking of food in a restaurant using an industrial robot and serving the food using a service robot, but someone still needs to put the food from the cooking robot on the plates and move the ready plates on the tray of the waiter robot. These tasks might be too few and too simple for human employees, thereby demotivating them. The company would be forced to combine these tasks with other unrelated tasks to create an economically feasible job position. However, the overall efficiency of this partially automated process and the attractiveness of the job position might suffer due to employees' inefficiency and demotivation to perform the unautomated tasks. Full automation eliminates the human operator and reorganises the processes to be performed by robots, AI and other automation technologies rather than human employees. On the one hand, this makes the service process less flexible, but, on the other hand, it improves its overall efficiency because all tasks will be automated. Customers' resistance to full automation due to lack of trust in or knowledge of how to use the technology would be gradually overcome when they become more exposed to fully automated tourism and hospitality companies with fewer service failures. Therefore, in the long run, the economic stimuli (decreasing technology costs, increased technology productivity and labour costs) work in favour of full automation in tourism and hospitality provided the availability of appropriate technological solutions.

One of the major issues on the road to full automation in tourism and hospitality is automated decision-making, elaborated on in Chapter 1. Automation is mainly associated with the implementation of tasks in front-of-house and back-of-house processes that are directly linked to service delivery (i.e. the primary activities in the terminology of Porter's [1985] Value Chain Model), such as cleaning, cooking, housekeeping, providing information, room service delivery, processing payments and documents, but not the managerial decisions related to these tasks. However, the automated tourism and hospitality company of the future will rely on automated decision-making as well. This means that artificial autonomous agents will make decisions related to prices, confirmation of bookings, choice of suppliers, market positioning, communication strategies, social media and metaverse posts, investment and financial decisions, etc., without human intervention (i.e. 'human-on-the-loop' and 'human-out-of-the-loop' approaches). The delegation of complete decision-making authority to artificial autonomous agents will make fully automated tourism and hospitality companies largely independent of human control. Human intervention, where necessary, would be related to physical repair whenever it cannot be automated. Additionally, humans will interfere in decisions that require significant emotional intelligence (e.g. dealing with complaints). Naturally, in a robonomic economy not all tourism and hospitality companies will be fully automated, and companies with various degrees of partial automation offering diverse experiences will successfully operate in the market.

6.4 Business Processes in the Automated Tourism and Hospitality Company of the Future

6.4.1 Operations management

Tourism and hospitality can be seen as a marketplace for experiences (Björk *et al.*, 2021). With the introduction of the experience economy (Pine & Gilmore, 1998, 1999, 2011), staging of experiences came to the fore as a way of mass-customising operations management to meet individuals' search for experiences (Loef *et al.*, 2017). Chapter 7 sheds more light on the implementation of automation technologies for creating tourism experiences. The co-creation of experiences (Campos *et al.*, 2018; Prahalad & Ramaswamy, 2004) is an evident part of tourism and hospitality (Moyle *et al.*, 2017). Hence, in the automated tourism company of the future, the design of experiences involving AI is key to operations management. New technologies pave the way for new types of experiences and novel ways of experiencing. These changes represent the automated experiencescape of the future.

Experiencescapes can be understood as 'nested products of inputs from organizations and tourists [...] produced through substantive and communicative staging' (Mossberg, 2007: 63), and extend beyond the servicescape (Bitner, 1992; Fossgard & Fredman, 2019; Vespestad & Hansen, 2019). It is important for automated tourism and hospitality companies to acknowledge that they are not limited to producing or delivering services to tourists. Operations management also involves a broader co-creation among several actors in an experience environment (Mossberg, 2007). In the future, some or all of these actors will be AI. This means that companies must take measures on how to maintain central aspects of co-creation (Aluri *et al.*, 2019), where the joint resources of the different actors are integrated into the experience (Prebensen *et al.*, 2014; Prebensen & Rosengren, 2016) with AI alleviating this process. This means that, for example storytelling (Pera & Viglia, 2016), immersion (Carù & Cova, 2006; Hansen & Mossberg, 2013) and even transformation (Lindberg & Østergaard, 2015) could be operated by AI. Moreover, AI can be integrated as a part of the experience in a way that facilitates a deeper immersion into the experience. For example, AI can be used pre-experience to lead tourists into the experience, providing facts for tourists to gain more knowledge about the Northern Lights (*Aurora Borealis*), or a robot could be used to tell myths about the Northern Lights to create a sense of magic.

Many small restaurants and capsule hotels will rely on (nearly) complete automation of tasks, thus effectively replacing labour as a production factor. From a marketing perspective, automated tourism experiences would be sold at low prices because automation technologies are largely perceived by tourists as cost-saving technologies, and tourists would wish that some of tourism and hospitality companies' cost savings due to automation be transferred to them through low prices (Ivanov & Webster, 2022). Additionally, the massive spread of automation in all sectors of the economy and all facets of social life and the associated economies of scale will lead to the commodification and standardisation of services, thus providing additional justification for low-priced automated experiences. At the same time, 'high-touch' tourism and hospitality companies may prefer to use human employees in their front-of-house operations, although some or all back-of-house operations might be automated. Their market positioning will be based on the human warmth in the service delivery process that technology cannot provide. Consequently, these 'high-touch' tourism and hospitality companies will charge higher prices than those that offer the automated experience. In between these two large groups of companies, there would be various shades of grey. Most tourism and hospitality companies will use a combination of human employees and automation to create tourism experiences. Thus, tourists will enjoy a constellation of options and will select the one that best fits them in terms of affordability and preferences towards the service provider (a human or a [ro]bot).

Previous studies have shown that tourists are not uniform in their attitudes towards robots and other automation technologies (Webster & Ivanov, 2021). They have different perceptions towards which tasks are appropriate for automation/robotisation and have different preferences towards the degree of automation of tourism and hospitality services. Hence, tourism and hospitality companies would need to identify the appropriate market segments for their products and to design the experiences based on tourists' preferences (Seyitoğlu & Ivanov, 2020). In the robonomic economy of the future, where automation is the norm, the provision of human-delivered and digital detox experiences will be the competitive advantage for some tourism and hospitality companies.

Going beyond the current economic paradigm and taking into consideration sustainability as a prevailing tourism management goal, operations management decisions will be made with the UN sustainable development goals (SDGs) as the gold standard. For example, pre-programmed questions relating to the contribution of operations to the fulfilment of the separate SDGs will be part of the daily procedures of both human employees and artificial autonomous agents. This would make the more sustainable options the default for operations management. This will apply to both internal and external activities, allowing for sustainably more favourable decisions to be made the benchmark. Chapter 10 sheds more light on the sustainability of tourism and hospitality in robonomics.

A major operational aspect of the automated tourism and hospitality company is the robot-friendliness of its facilities (Ivanov & Webster, 2017a). It relates to how the design and condition of the facilities support a mobile robot to fulfil its tasks (e.g. to deliver a room service order). The robot-friendliness of hospitality facilities depends on the presence of doors, doorsteps, floor inclination, tidiness, the presence or lack of artificial landmarks and sensors to help with robot's navigation, etc. The more robot-friendly the facilities of a hospitality company, the easier the implementation of service robots in it. Considering that many of the requirements for robot-friendly facilities are the same as for wheelchair access (e.g. lack of steps, floor surface inclination), more and more hospitality companies in the future will have robot-friendly facilities that contribute towards the successful implementation of service robots in their operations. While this applies to built facilities such as theme parks, hotels and museums, for outdoors nature-based tourism, particularly those with an experiential focus, robot friendliness is not evident to the same extent. Chapter 8 delves deeper into the development of robot-friendly hospitality facilities.

6.4.2 Marketing management

The automated tourism company of the future will use artificial autonomous agents to make marketing-related decisions about its target customers, market positioning, determining the elements of the marketing mix, marketing communication campaigns, budgeting, communicating with customers, etc. At the same time, tourists will use their own agents to find and compare tourist offers and to book tourist services for them. This means that on the market the artificial autonomous agents of tourism and hospitality companies will face the artificial autonomous agents of customers. Therefore, companies' autonomous agents will need to target the autonomous agents of customers, i.e. to apply AI2AI marketing (Ivanov, 2022). The concept of AI2AI marketing acknowledges that both tourism supply and demand may utilise AI. Therefore, tourism companies need to redesign their marketing activities to reflect the fact that purchase decisions might not be taken by humans but by AI. Autonomous agents will be able to quickly evaluate tourist offers, book, cancel and rebook services as prices change on the market. Besides the functional characteristics of the offers (e.g. prices, trip duration and flight schedule), the autonomous agents might consider other characteristics as well, such as the sustainability of operations, the carbon footprint and the fair trade practices of suppliers, which are currently nearly impossible to assess but the abundance of data in robonomics might make it feasible. Therefore, the marketing management of tourism companies in the future needs to adopt a broader perspective of the criteria that customers would use in their purchase decisions. From an automated marketing perspective, focusing on the interaction between businesses and consumers, using AI will be essential in building tourism brands. The ability of AI to build expectations and to follow up customers, e.g. after a visit to a tourist attraction or a themed hotel, represents mass-customisation in a way that can prove both economically viable and initiate a revisit intention.

Moreover, the definition of a customer will change because from an accounting perspective, the outcomes of a purchase decided and implemented by AI are indistinguishable from those of a purchase decided and implemented by a human. Therefore, the artificial autonomous agents that make purchase decisions need to be considered as customers as well (Ivanov & Webster, 2017b). Moreover, robots may become users of some tourist services (e.g. visits to museums and galleries, city tours) (Ivanov, 2019b). Future robots will be quite sophisticated in terms of emotional intelligence; they will be able to perceive, use, understand and manage the emotions of others, but it is not yet clear whether they will continue to have the inability to experience emotions. Considering that emotions largely shape tourism experiences (Kim & Fesenmaier, 2015; Volo, 2017), it is difficult to predict robots'/artificial autonomous agents' level of emotional engagement with and the ability to appreciate the tourism services they use. Additionally, human employees of tourism and hospitality companies may resist serving robot customers and consider them as inferior entities, an issue that will not exist in a fully automated company.

The delegation of decision-making authority to artificial autonomous agents in the area of tourism and hospitality marketing will lead to other implications as well (Ivanov, 2022). First, the information asymmetry between buyers and sellers in the tourism market will decrease because both sides will use AI and process large amounts of data. Second, there will, however, be information asymmetry between the artificial autonomous agents and their human/organisational owners – agents will have more information about their decisions and actions compared to humans. Human customers may doubt the decisions of their own agents. Therefore, tourism and hospitality companies need to develop marketing campaigns targeted at human customers with the message that their agents made the right decisions to avoid any cognitive dissonance. Third, prices will be more stable because the agents will not have stimuli to change them. For instance, if a seller artificial autonomous agent increases the price due to the decreased information asymmetry between buyers and sellers, the buyer agents will quickly detect this and direct purchases to other seller agents that did not increase their prices. Similarly, if a seller agent lowers the price, the other seller agents will quickly detect the move and lower their prices, thus eliminating any temporary price advantage and the stimuli to lower prices. Fourth, some autonomous agents may conceal or share incorrect information about the identity and characteristics of their owners in order to obtain better prices and conditions. Therefore, researching the characteristics of potential human customers might be challenging. Fifth, tourism companies might be in a 'war of the defaults settings' (Webster, 2022). They will compete on whose apps to preinstall in smart speakers, smartphones, tablets and other devices to gain a competitive advantage over other companies. Sixth, due to the automation of marketing activities and big data analytics, the marketing communications of tourism and hospitality companies will be omni-device, omni-channel and hyper-personalised to individual human customers. They will also be less aggressive and more subtle and personalised, e.g. a hotel in a movie may appear with a different logo and brand name to different viewers and even to the same viewers who watch it at different times. This will be further facilitated by human microchip implants for human augmentation that might be widespread among tourists (Ivanov *et al.*, 2014).

6.4.3 Financial management

The artificial autonomous agents running the automated tourism and hospitality companies will have real-time information about their companies' assets, liabilities, revenues, costs, cash inflows and outflows, and cash needs. They will forecast with a high degree of accuracy their cash inflows (e.g. sales) and outflows (e.g. payments to suppliers), assess their cash needs or excess cash, quickly evaluate the available investment opportunities and current bank loan offers, direct their excess cash to short-term investments or take loans at the best available terms whenever they need them, often for only a few hours.

The fully automated tourism and hospitality company will not have human employees, thus there will be no payments for salaries, social security and health insurance. The lack of labour-related payments significantly simplifies the financial management because it eliminates a large, fixed cost whose payment is strictly regulated by legislation. However, there will be payments to suppliers (often other automated companies in or outside tourism), insurances, taxes and payments to shareholders. The technologisation of service delivery in tourism and hospitality means that automated companies will have a lot of assets in the form of physical equipment that is subject to depreciation, maintenance and insurance that are fixed costs. The share of fixed costs in the total costs of tourism and hospitality companies in robonomics will likely decrease because in this currently labour-intensive industry, labour-related costs may sometimes account for nearly half of total costs (Geffroy, 2019). The increase in the technology-related costs due to the use of more technological solutions will be lower than the decrease in the labour costs because otherwise, the implementation of automation would not make economic sense if it increases the total costs (see Ivanov & Webster [2019] for a more detailed elaboration of the economic fundamentals of the use of automation technologies in tourism and hospitality). The decreased share of fixed costs in the total costs means that automated tourism and hospitality companies will have a lower break-even point compared to companies without automation.

6.4.4 Legal issues

A major consideration related to automated tourism and hospitality companies in the future is their rights and obligations. Currently, companies may be owned by physical persons or other legal entities (companies, organisations, foundations, public authorities, etc.), but in robonomics artificial autonomous agents will own assets like any other legal entity, including shares in tourism and hospitality companies. Fully automated tourism and hospitality companies owned and run by AI will have the same rights and obligations as tourism and hospitality companies owned by humans and legal entities created or owned by humans – e.g. to pay taxes, to enter into contractual relationships with tourists, suppliers and distributors and fulfil their contractual obligations to them, to adhere to legal requirements about the issue and renewal of licences (e.g. for selling alcohol) the categorisation of premises and the maintenance and repair of facilities. To fulfil the legal obligations of the companies they own and run, in the future, legislation will need to grant some limited economic rights to artificial autonomous agents to allow them to own tourism and hospitality companies, to have bank accounts, to make and receive payments, to sign contracts, to make investment decisions, to pay taxes, etc. (see Chapter 3). In doing so, the artificial autonomous agents will become fully fledged participants in the robonomic economy and will be able to run fully automated tourism and hospitality companies without human intervention because they will be able to make decisions that are currently reserved for humans. At the same time, giving economic rights to artificial autonomous agents needs to be accompanied by legal obligations – e.g. pay taxes, comply with the laws and respond to tourist complaints. An important component of legislation will be related to liabilities for damages caused by autonomous agents (Hacker, 2023) and associated compulsory insurances for such damages for all tourism and hospitality companies that use robots and AI that could somehow cause damage to the property, to the health and life of tourists, employees, the company or third parties. The drive for such compulsory insurances may come from national legislation, supranational organisations (e.g. the European Union) or be mandated by insurance companies in the insurance packages they provide to tourism and hospitality companies.

6.5 Conclusion

Automation is an integral part of the future of tourism and hospitality. Companies will use robots, AI and various other automation technologies to co-create experiences for/with tourists and automate back-of-house and front-of-house processes. Many companies will completely automate all processes, including decision-making. They will offer affordable 'high-tech' experiences to their customers. Some of these companies will even be owned by artificial autonomous agents rather than humans. Other companies will not automate front-of-house processes and will deliver expensive 'high-tech' experiences to guests, although they will likely automate many of their back-of-house processes. Between these two extremes, tourism and hospitality companies will employ various degrees of partial automation (Ivanov, 2019a). Human-delivered tourism services would be the exception rather than the norm and together with digital detox vacations they will offer an escape from the technology-overloaded daily routine.

The future customer will not necessarily be a human, but purchase decisions will mostly be taken by artificial autonomous agents. Therefore, the scope of the marketing activities of companies will broaden to focus on AI customers as well. Moreover, robots may be tourists themselves (Ivanov, 2019b), although it is not yet clear whether they will have emotional engagement with the tourism activities they participate in or they will only mimic what humans do. Ultimately, during robonomics, tourism will cease to be a 'people's business' because machines will be serving machines, but it won't be considered perverse.

References

Aluri, A., Price, B.S. and McIntyre, N.H. (2019) Using machine learning to cocreate value through dynamic customer engagement in a brand loyalty program. *Journal of Hospitality & Tourism Research* 43 (1), 78–100. https://doi.org/10.1177/1096348017753521

Bainbridge, L. (1983) Ironies of automation. *Automatica* 19 (6), 775–779.

Bitner, M.J. (1992) Servicescapes: The impact of physical surroundings on customers and employees. *Journal of Marketing* 56 (2), 57–71.

Björk, P., Prebensen, N., Räikkönen, J. and Sundbo, J. (2021) 20 years of Nordic tourism experience research: A review and future research agenda. *Scandinavian Journal of Hospitality and Tourism* 21 (1), 26–36. https://doi.org/10.1080/15022250.2020.1857302

Bornet, P., Barkin, I. and Wirtz, J. (2021) *Intelligent Automation: Welcome to the World of Hyperautomation*. World Scientific Publishing Company.

Campos, A.C., Mendes, J., Valle, P.O.d. and Scott, N. (2018) Co-creation of tourist experiences: A literature review. *Current Issues in Tourism* 21 (4), 369–400. https://doi.org/10.1080/13683500.2015.1081158

Carù, A. and Cova, B. (2006) How to facilitate immersion in a consumption experience: Appropriation operations and service elements. *Journal of Consumer Behavior* 5 (1), 4–16.

Carvalho, I. and Ivanov, S. (2024) ChatGPT for tourism: Applications, benefits, and risks. *Tourism Review* 79 (2), 290–303. https://doi.org/10.1108/TR-02-2023-0088

Fossgard, K. and Fredman, P. (2019) Dimensions in the nature-based tourism experiencescape: An explorative analysis. *Journal of Outdoor Recreation and Tourism* 28, 100219. https://doi.org/10.1016/j.jort.2019.04.001

Geffroy, L. (2019, November 18) Higher wages are tied to higher efficiency, data shows. *CoStar*. https://www.costar.com/article/110751413/higher-wages-are-tied-to-higher-efficiency-data-shows (accessed 15 October 2023).

Hacker, P. (2023) The European AI liability directives: Critique of a half-hearted approach and lessons for the future. *Computer Law & Security Review* 51, 105871. https://doi.org/10.1016/j.clsr.2023.105871

Hansen, A.H. and Mossberg, L. (2013) Consumer immersion: A key to extraordinary experiences. In J. Sundbo and F. Sørensen (eds) *Handbook on the Experience Economy* (pp. 209–227). Edward Elgar.

Iskender, A. (2023) Holy or unholy? Interview with open AI's ChatGPT. *European Journal of Tourism Research* 34, 3414. https://doi.org/10.54055/ejtr.v34i.3169

Ivanov, S. (2019a) Ultimate transformation: How will automation technologies disrupt the travel, tourism and hospitality industries? *Zeitschrift für Tourismuswissenschaft* 11 (1), 25–43.

Ivanov, S. (2019b) Tourism beyond humans: Robots, pets and teddy bears. In G. Rafailova and S. Marinov (eds) *Tourism and Intercultural Communication and Innovations* (pp. 12–30). Cambridge Scholars Publishing.

Ivanov, S. (2020) The impact of automation on tourism and hospitality jobs. *Information Technology & Tourism* 22 (2), 205–215. https://doi.org/10.1007/s40558-020-00175-1

Ivanov, S. (2022) AI2AI marketing: Foundations and research agenda. *ROBONOMICS: The Journal of the Automated Economy* 3, 26. https://journal.robonomics.science/index.php/rj/article/view/26 (accessed 19 December 2022).

Ivanov, S. (2023) The economics of generative AI. In Z. Lv (ed.) *Applications of Generative AI*. Springer Nature (in press)

Ivanov, S. and Webster, C. (2017a) Designing robot-friendly hospitality facilities. Proceedings of the Scientific Conference 'Tourism. Innovations. Strategies', 13–14 October, Bourgas, Bulgaria, pp. 74–81.

Ivanov, S. and Webster, C. (2017b) The robot as a consumer: A research agenda. Proceedings of the 'Marketing: Experience and Perspectives' Conference, 29–30 June, University of Economics-Varna, Bulgaria, pp. 71–79.

Ivanov, S. and Webster, C. (2019) Economic fundamentals of the use of robots, artificial intelligence and service automation in travel, tourism and hospitality. In S. Ivanov and C. Webster (eds) *Robots, Artificial Intelligence and Service Automation in Travel, Tourism and Hospitality* (pp. 39–55). Emerald Publishing. https://doi.org/10.1108/978-1-78756-687-320191002

Ivanov, S., Webster, C. and Mladenovic, A. (2014) The microchipped tourist: Implications for European tourism. In A. Postma, J. Oskam and I. Yeoman (eds) *The Future of European Tourism* (pp. 86–106). Stenden University of Applied Sciences.

Ivanova, M., Jeliazkova, E. and Ivanov, S. (2021) Customised vs. standardised hotel service: Are customers willing to pay more for a customised experience? *Yearbook of Varna University of Management* 14, 88–98.

Kim, J. and Fesenmaier, D.R. (2015) Measuring emotions in real time: Implications for tourism experience design. *Journal of Travel Research* 54 (4), 419–429.

Lindberg, F. and Østergaard, P. (2015) Extraordinary consumer experiences: Why immersion and transformation cause trouble. *Journal of Consumer Behaviour* 14 (4), 248–260. https://doi.org/10.1002/cb.1516

Liu, P. and Li, Z. (2012) Task complexity: A review and conceptualization framework. *International Journal of Industrial Ergonomics* 42 (6), 553–568. https://doi.org/10.1016/j.ergon.2012.09.001

Loef, J., Pine, B.J.I. and Robben, H. (2017) Co-creating customization: Collaborating with customers to deliver individualized value. *Strategy & Leadership* 45 (3), 10–15. https://doi.org/10.1108/SL-03-2017-0028.

Mossberg, L. (2007) A marketing approach to the tourist experience. *Scandinavian Journal of Hospitality and Tourism* 7 (1), 59–74.

Moyle, B.D., Scherrer, P., Weiler, B., Wilson, E., Caldicott, R. and Nielsen, N. (2017) Assessing preferences of potential visitors for nature-based experiences in protected areas. *Tourism Management* 62, 29–41. https://doi.org/10.1016/j.tourman.2017.03.010

Pera, R. and Viglia, G. (2016) Exploring how video digital storytelling builds relationship experiences. *Psychology & Marketing* 33 (12), 1142–1150. https://doi.org/10.1002/mar.20951

Pine, B.J. and Gilmore, J.H. (1998) The experience economy. *Harvard Business Review* 76 (6), 97–105.

Pine, B.J. and Gilmore, J.H. (1999) *The Experience Economy, Work Is Theatre & Every Business a Stage*. Harvard Business School Press.

Pine, B.J. and Gilmore, J.H. (2011) *The Experience Economy*. Harvard Business Review Press.

Porter, M.E. (1985) *Competitive Advantage. Creating and Sustaining Superior Performance*. The Free Press. Reprinted in 1998 with a new introduction.

Prahalad, C.K. and Ramaswamy, V. (2004) Co-creation experiences: The next practice in value creation. *Journal of Interactive Marketing* 18 (3), 5–14.

Prebensen, N.K. and Rosengren, S. (2016) Experience value as a function of hedonic and utilitarian dominant services. *International Journal of Contemporary Hospitality Management* 28 (1), 113–135. https://doi.org/10.1108/IJCHM-02-2014-0073

Prebensen, N.K., Chen, J.S. and Uysall, M.S. (eds) (2014) *Creating Experience Value in Tourism*. CABI.

Seyitoğlu, F. (2021) Automation vs authenticity in services. *ROBONOMICS: The Journal of the Automated Economy*, 2, 20. https://journal.robonomics.science/index.php/rj/article/view/20 (accessed 16 December 2021).

Seyitoğlu, F. and Ivanov, S. (2020) A conceptual framework of the service delivery system design for hospitality firms in the (post-)viral world: The role of service robots. *International Journal of Hospitality Management* 91, 102661. https://doi.org/10.1016/j.ijhm.2020.102661

Slack, N., Brandon-Jones, A. and Burgess, N. (2022) *Operations Management* (10th edn). Pearson

Tuomi, A., Tussyadiah, I.P. and Stienmetz, J. (2021) Applications and implications of service robots in hospitality. *Cornell Hospitality Quarterly* 62 (2), 232–247. https://doi.org/10.1177/1938965520923961

Vespestad, M.K. and Hansen, O.B. (2020) Shaping climbers' experiencescapes: The influence of history on the climbing experience. *Journal of Hospitality & Tourism Research* 44 (1), 109–133. https://doi.org/10.1177/1096348019883685

Volo, S. (2017) Emotions in tourism: From exploration to design. In D.R. Fesenmaier and Z. Xiang (eds) *Design Science in Tourism: Foundations of Destination Management* (pp. 31–40). Springer.

Webster, C. (2022) War of the default settings. *ROBONOMICS: The Journal of the Automated Economy* 3, 38. https://journal.robonomics.science/index.php/rj/article/view/38 (accessed 19 December 2022).

Webster, C. and Ivanov, S. (2021) Attitudes towards robots as transformational agents in tourism and hospitality: Robophobes vs. robophiles. In A. Farmaki and N. Pappas (eds) *Emerging Transformations in Tourism and Hospitality* (pp. 68–82). Routledge.

7 Creating Experiences through Automation Technologies

Katerina Berezina, Lisa Cain, Katerina Volchek and Cihan Cobanoglu

7.1 Introduction

The speed of revolutionary transformations that automation brings to everyday life will likely affect the travel experience of the future. The opening scenario in the first chapters of this book presents a hospitality and tourism technology–customer experience from start to finish in 2068. Grounded in current theoretical foundations and technological advancements, this scenario shares a vision of how the development and integration of different automation technologies may alter and shape the travel experience in the state of robonomics, or the robot-dominated economy (Ivanov, 2017). The scenario envisions a couple of travellers who use a virtual assistant to search for and book a trip according to personal preferences and budgetary constraints (Yanishevskaya *et al.*, 2019); benefit from the joint efforts of a virtual assistant and autonomous vehicles to be reminded of the appropriate departure time to arrive at the airport on time (Cohen & Hopkins, 2019); employ biometric technology for precise and speedy traveller recognition facilitating seamless passage through airport security (Negri *et al.*, 2019); use mobile technology to check in to the hotel at a requested time; and enjoy personalised services offered to them based on past consumption history and guest preferences learned from the consumer's behaviour.

This chapter revisits the dimensions of the service experience to adjust them to the realities of robonomics, highlighted by Tussyadiah (2020) and Xiang *et al.* (2021). The chapter then re-examines the sequence of service and the technological points of contact in the service exchange. It begins with examining the service selection stage (Choose), including a review of booking options and capabilities, takes the reader through the travel experience (Go) to the check-in and stay experience (Live), as well as dining (Eat) and fitness (Exercise) experiences, and culminates with post-departure follow-up technology (Review). By reviewing innovations brought about by robotics and automation at every step of the travel process, the chapter envisages in detail how robots, autonomous vehicles and devices, artificial intelligence (AI) and technology will create a positive experience for the customer. It also discusses how the robotic service recovery will transpire. Finally, the chapter examines negative robotic experiences and how single or multiple destructive experiences will influence satisfaction and loyalty.

7.2 Theoretical Foundations

The current literature often considers high tech as the opposite of high touch, suggesting that technology may be detrimental to customer experiences (Neuhofer *et al.*, 2014). Based on the stimulus–organism–response (SOR) framework (Mehrabian & Russell, 1974), this chapter suggests that as the economy transitions to the state of robonomics, consumers will more frequently experience interactions with robots (S). As the robotic stimuli change, robots become more efficient and accurate and occur more frequently (Ivanov, 2020), the organism's emotional responses (O) will evolve as well, changing consumer behavioural responses (R) to robot-created experiences. Based on this logic, the chapter recognises the need to evolve our understanding of the service experience to include technology-enabled experiences.

To imagine the possible nature of automation-enabled services and the resulting travel experience, this chapter builds on the process of innovation through co-creation (Helkkula *et al.*, 2018). According to co-creation, service is a process of resource integration conducted by several stakeholders to create value for each of them (Vargo *et al.*, 2020a). Thus, the success of this process depends on the targeted value (e.g. advanced tourist experience), integrated resources (e.g. technology and data, knowledge and skills) and defined processes for resource integration (e.g. methods of tourist data collection, methods of cleaning hotel rooms, principle of personalising a service for the tourist), required to deliver the desired service outcome (Vargo *et al.*, 2020b). To target advanced travel service through innovation, travel businesses need to either redefine their output (i.e. desired value and related service parameters) or optimise their applied resources and the processes of their integration by the stakeholders.

The complexity in delivering an innovative travel service is determined by the nature of the tourist experience. In addition to purely utilitarian needs, tourists are motivated to acquire experiential value (McKercher, 2016). Experience can be gained when tourists are immersed in interactions with a travel service and perceive these interactions as entertaining, educational and aesthetical, enabling them to escape from the burden of everyday reality (Pine & Gilmore, 1999). Importantly, the proliferation of information and communication technologies and, recently, of robotics and automations, in everyday life, has raised tourist expectations (Buhalis & Sinarta, 2019). Therefore, innovative service processes and output do not guarantee advanced tourist experience. To deliver value, a business strategy would plan for innovative, highly personalised tourist experiences.

The complexity of designing an advanced tourist experience in smart environments requires a holistic approach (Xiang *et al.*, 2021). Such an approach would strategise an innovation ecosystem with its stakeholders, resources, processes, planned outcomes and experiences (Shin *et al.*, 2019). The systemic strategy of innovation helps to ensure that value is co-created for all stakeholders. To envision the future transformation of the tourist experience by automation and robotics in the state of robonomics, this chapter highlights the process of experiential value co-creation from the perspective of each of four innovation archetypes: process based, output based, experiential innovation and systemic innovation (Helkkula *et al.*, 2018; Shin *et al.*, 2019).

Process-based innovation focuses on the process of designing and delivering the service experience and may include changes in service elements and their sequencing, as well as the roles and skills of customers and employees. The output-based innovation archetype deals with the quantifiable results (outputs) of service transformation. This innovation type considers measurable outcomes that may benefit all stakeholders (i.e. customers, employees and businesses). Customer individual experiences and perceptions of service and its value are at the core of the experiential innovation archetype. While literature documents various approaches to assessing customer experience and value perception of services, this chapter focuses on the approach proposed by Pine and Gilmore (1999) who suggested that optimal experience effects are achieved when the customer experiences absorption, active and/or passive participation and immersion in the experience. These types of interactions are achieved through entertainment, education, aesthetics and escapism. And, the systemic innovation archetype looks at the redesigned service experience holistically (Helkkula *et al.*, 2018; Shin *et al.*, 2019). This approach zooms out from a specific technology or change that was introduced to drive innovation and looks at the service as an ecosystem where the system-wide value is larger than the sum of all system elements.

Given the major shift that all hospitality businesses will undergo to achieve robonomics, the chapter proposes that all businesses will need to re-evaluate their processes to achieve efficiency in their output and success in creating satisfying experiences for the customers. Therefore, the evaluation of future automation-enhanced travel experiences is conducted using a proposed framework for Creating Rich Experiences through Automation Technology, Innovation and Virtual Environments (CREATIVE). The CREATIVE framework (Figure 7.1) has innovation at its core and impacts three major elements of the service process: the service selection, core service and service evaluation stages. This

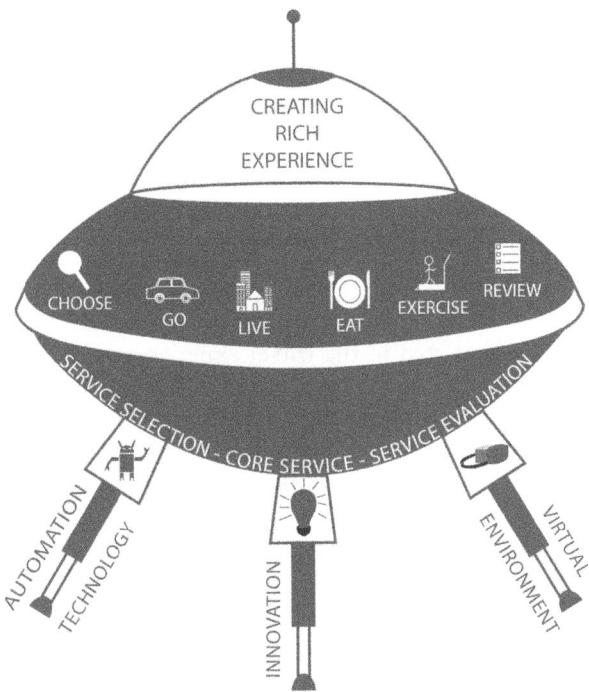

Figure 7.1 The Creating Rich Experiences through Automation Technology, Innovation and Virtual Environments (CREATIVE) framework

framework may be customised to fit the needs of different service sectors, such as CREATIVE healthcare, CREATIVE education or CREATIVE entertainment. The CREATIVE hospitality framework expands the initial three elements to include sub-elements specific to the hospitality and tourism industry. For example, the service selection stage corresponds to the choose step in the travel experience, the core service is broken down into go, live, eat and exercise experiences, and the service evaluation stage is conceptualised as the review step of the travel process.

As the core force in the CREATIVE framework, innovation is altering the travel experience by driving changes at the process, output, experience and system levels (Helkkula et al., 2018; Shin et al., 2019). Process-based innovation in this case refers to changes in the process of creating hospitality and travel experiences. This may be achieved by modifying the roles of service providers and travellers in service delivery, such as altering the required skills and incorporating new technologies by all involved parties (Ramaswamy & Ozcan, 2018). Output-based innovation describes the improvement in the delivered service characteristics to enhance various areas. For example, financial effectiveness may bring increased profitability to an organisation and cost savings to travellers. Improved visual design of a destination website may aim to bring aesthetical pleasure and support travellers in their decision-making (Ivanov, 2020). Experience-based innovation reveals itself through customer perception of the offered experiential value of the hospitality and tourism product (Neuhofer *et al.*, 2014). And, finally, system-level innovation speaks to the improvement of the underlying system, resources and infrastructure that support service creation (Xiang *et al.*, 2021). Next, the chapter considers the impacts of the four types of innovation across all elements of hospitality and travel experience.

7.3 Choose

Choose is the first step in the travel experience that corresponds to the service selection element in the CREATIVE framework. At this step, customers start dreaming about their vacations, consider potential alternatives and make their choices (Buhalis & Volchek, 2021). Our discussion of the CREATIVE framework begins with considering the impacts of innovation on the way customers will *choose* their trips in the state of robonomics because this is the step that starts the entire experience.

7.3.1 Travel needs and trip planning

The complexity of creating advanced tourists' experiences is directly related to the variable nature of tourist needs and demands (Volchek *et al.*, 2021). Using the broadest lens, tourist motives to travel have remained the same and are likely to continue to drive customers to travel to relax, recover from stress and everyday routines, as well as to acquire new, authentic and memorable experiences (Pearce, 2011). Recent trends demonstrate the increased demand for individualisation, personalisation and engagement in the co-creation of experiences (Buhalis & Sinarta, 2019). At the same time, the development of smart technologies and ambient intelligence will likely increase tourist expectations of services provided by real-time automation and the seamless experience of need satisfaction (Buhalis, 2019; Yang *et al.*, 2021). Tourism providers of the future will face the challenge of supporting travel planning with highly personalised services that have the potential to deliver advanced experiences in a way that is non-intrusive and does not create additional cognitive and emotional load (Volchek *et al.*, 2020).

7.3.2 Process-based innovation

The spread of smart environments with embedded sensors, interconnected devices and advanced analytics will create unprecedented opportunities to identify immediate customer needs (Tussyadiah, 2020). Today, travel planning explores the opportunities to leverage consumer data, such as customers' previous trips, search histories, locations, available times and budgets, as well as destination data on events, in order to understand individual tourist preferences (Volchek *et al.*, 2020). The proliferation of portable sensors and the advancements of automation in data collection and analytics are expected to boost neuromarketing (Moyle *et al.*, 2019; Prentice, 2019). The incorporation of real-time scans of tourist physiological parameters by smart wearables and, soon, by implanted microchips, will be continuously observed by virtual assistants to define the state of health, mood, level of stress, as well as immediate cognitive and emotional reaction to proposed services (Ivanov *et al.*, 2014; Mercan *et al.*, 2020) This will advance tourist need recognition by enabling virtual assistants to define actual demand for travel and the need for specific services in real time.

7.3.3 Output-based innovation

The capability of understanding the tourist demand for travel will enable the service design process to produce radically new service offerings. Relevant accommodation, transportation, dining, excursions and other services will be automatically selected to create highly personalised travel offers for each tourist. Virtual assistants (Tung & Law, 2017) will then take care of delivering such offers to the relevant place in a real or digital environment, and at the moment of the travel need recognition or even before it. As a result, automated virtual assistants will create a space of completely new types of personalised services, capable of delivering advanced experiences.

7.3.4 Experience-based innovation

Data-driven personalisation is named among the determinants of advanced tourist experiences (Buhalis & Sinarta, 2019). Relevant services, delivered prior to the moment when the tourist starts an extensive information search, can eliminate the 'search' stage. This will allow the tourist to make a seamless transition from need recognition to the decision stage of the customer journey without investing time in planning activities (Buhalis & Volchek, 2021). Such transformative innovation may have a two-fold effect on tourist planning experience. First, rather than simply decreasing the cognitive and emotional loads of decision-making, it will eliminate them by default (Ivanov & Webster, 2017). Second, the decision-making phase constitutes the events that form the tourist experience (Neuhofer *et al.*, 2014). The elimination of information overload will enable tourists to redirect their time and effort on a travel information search and analysis to gett inspiration from the proposed services and build expectations of the future trip. As a result, advanced customer experience of the trip planning stage can be created through service automation.

7.3.5 Systemic innovation

The capabilities of the service automation of customer context recognition and personalisation, conceptualised above, represent the systemic innovation of customer need recognition. The ability of the service ecosystem to co-create an advanced tourist experience depends on the availability of all resources, including the requisite data, technology and skills to apply them, as well as on the willingness of all stakeholders to contribute them in a way that is determined by the system (Ramaswamy & Ozcan, 2018). In other words, outdated data about tourists' needs, incorrect inference of at least one of the automation processes, including data cleaning, pattern recognition, demand interpretation and service personalisation, may result in an irrelevant service and dissatisfaction (Volchek *et al.*, 2020). This creates the risk of co-destruction for the tourist planning experience, as well as for the whole trip. Therefore, a comprehensive and dynamic approach for managing the systemic innovation of the future will become even more important than it is today.

7.4 Go

Following the service selection element, the CREATIVE framework focuses on the core service. The core service of the travel experience may be broken down into several steps that include go, live, eat and exercise. This chapter continues to review the *go* step of the travel experience that describes the ways in which customers will reach their destinations when travelling.

7.4.1 Transportation advances driving change in travel experiences

The growth and development of automation technology will significantly change the way people travel. Such changes may include different types of self-driving vehicles, including airport shuttles, taxis, car rentals and sightseeing tours (Cohen & Hopkins, 2019; Globetrender, 2016; Ivanov *et al.*, 2017, 2020; Tussyadiah *et al.*, 2017), ultra-fast trains and aeroplanes. All of these advancements in the transportation landscape may alter customer preferences when they travel and expand the destination choice and potential reach for customers. Process-, output-, experience-based and systemic innovations brought about by these advancements are presented and discussed below.

7.4.2 Process-based innovation

The process-based change in the travel experience will begin with the introduction of a virtual assistant in the planning and organisation process (Tung & Law, 2017). Having access to the traveller's calendar, location, distance to the destination and traffic information, virtual assistants can take care of planning transfers and transportation. Knowing the car traffic conditions on the road, as well as the foot traffic at train stations and airports, a virtual assistant may dynamically adjust the departure time to ensure a smooth travel experience. Also, according to customer preferences, a virtual assistant may make the right choice of the type of transportation, selecting, e.g. a shared or private taxi ride, or a business or economy cabin class. Customer preferences paired with the information of biometric scans and travel time estimates may allow virtual assistants to select the fastest or a scenic route to further enhance the travel experience.

The process-based change in the travel part of the tourist experience is further driven by advancements in the transportation sector (Cohen & Hopkins, 2019). With increased automation in the travel experience, travel distance and transportation types will be revisited. Self-driving cars will become more comfortable, be able to travel longer distances and will take the need to drive out of the equation, thereby offering a convenient alternative to train travel. Ultra-fast trains may cover longer distances, thus making them viable competition for short-haul flights. Long-haul flights may be covered in shorter times, offering travellers an opportunity to visit remote destinations that were not previously available due to financial or time constraints (Prideaux, 2019).

Advancements in self-driving vehicle development may prompt societies to rethink owning a vehicle in favour of a shared model of car transportation (Cohen & Hopkins, 2019; Prideaux, 2019). This model may allow societies to minimise traffic and offer more affordable fares. The services may be offered in tiers to accommodate the needs of different types of customers (e.g. those who are or are not willing to travel with strangers due to family composition, cultural reasons or safety concerns) (Buhalis & Sinarta, 2019). The shared model of car transportation may blur and, finally, fully remove the lines between transportation for locals and tourists. The entire city infrastructure should be available to everyone and support the needs of both groups.

Automation and robotisation will inevitably bring changes to the transportation hub (train stations and airports) processes, including check-in, luggage handling, security and health scanning (Ioannou *et al.*, 2020; Tussyadiah, 2020). The commute becomes more efficient and convenient as self-driving vehicles drop passengers off at the transportation hub entrance and leave to self-park, thus eliminating the need to search for parking, park the vehicle and commute to the terminal. All processes within the transportation hub are streamlined using biometric corridors to identify passengers, register their presence in the facility and verify their travel details and visa privileges. Built-in radio-frequency identification (RFID) chips on each suitcase may be reprogrammed prior to luggage drop off to include the traveller destination for that trip (Mercan *et al.*, 2020). Biometric scans also contribute to enhanced security and health screenings making it easier to identify travellers who could be dangerous to other passengers. Finally, implanted microchips (Ivanov *et al.*, 2014) and biometric corridors would serve as reliable sources of the incoming foot traffic at the transportation hub, feeding this information back to virtual assistants who are planning traveller transfers (Tung & Law, 2017). Transportation hub automation should create more enjoyable, secure and time-efficient experiences for travellers by eliminating lines, manual traveller screenings and document checks.

Automation in all transportation sectors may eliminate human-to-human interaction between passengers and drivers (Cohen & Hopkins, 2019). If a shared ride is used, tourists could potentially engage with locals. If not, tourists may use their time to ask questions of their virtual assistants. Additionally, changes in the transportation process may provide opportunities for advertising. Similarly, beyond transportation, all aspects of the transportation hub experience will be affected by removing interaction with check-in and security agents, store attendants and restaurant wait staff from the travel experience (Ivanov & Webster, 2019). Transportation hubs may think of potential ways to showcase the culture and traditions of the region. Mini stages in different areas of the airport may offer an opportunity to employ humans who would perform local dances and songs or showcase crafts, and by doing so further improve the traveller experience.

7.4.3 Output-based innovation

Automation and robotisation of the travel experience may result in positive outputs, such as time, safety and financial improvements, enhanced travel efficiency and increased accessibility of travel (Tussyadiah, 2020). When it comes to time, customers no longer have to own a car but may utilise vehicles available for shared use that are effective, planned and timely. New modes of transportation have the potential to reduce travel times and increase time utility (Cohen & Hopkins, 2019). Self-driving cars and high-speed trains may serve as substitutes for short-haul planes (Prideaux, 2019). Advanced planes may cover the same distances in shorter times. The utility of the travel time improves because guests may sleep, eat or sightsee during the trip, or just have quality time with their loved ones while the driving, planning and coordination are taken care of by automation. Additionally, time benefit is gained from having all processes in the travel experience optimised by ID and individual preferences or restrictions recognition from implanted microchips that helps with removing lines, bureaucracy and wait.

Another output innovation in the travel experience is the new level of safety (Prideaux, 2019). Automation offers a system-controlled environment that eliminates human errors and their associated risks. Self-driving cars are expected to be safer than human-operated vehicles (Cohen & Hopkins, 2019). Thus, when implemented on a large scale, driverless cars may bring an overall improvement to the safety of the travel experience (Tussyadiah, 2020). Similarly, airports provide safer environments by conducting biometric scans of all passengers for personal identification, explosives, visa privileges and health conditions, e.g. by using microchip technology.

After initially charging a premium for innovation, self-driving vehicles are likely to offer a more economical way to travel by eliminating the driver, thus offering financial benefits to travellers (Cohen & Hopkins, 2019). Additionally, if the car ownership model is abandoned, the cost of owning and maintaining a vehicle is shared among all passengers through the rental price or the ticket. Overall, the cost of travel may be reduced by using self-driving cars to cover longer distances than driver-based vehicles. As other transportation sectors develop and become fully automated, cost savings will come through in those areas as well.

The efficiency of the travel experience is maximised by improving traffic flow (Cohen & Hopkins, 2019) as vehicles should be able to communicate with each other and the overarching infrastructure for finding better routes. As all modes of transportation become faster, the commute part of the travel experience may be shortened, allowing more time at the destination or it could be turned into an experience in itself for those seeking road trip type of travel (Gretzel, 2022). The efficiency of transportation hubs is achieved by automating and optimising all processes, e.g. removing the need for manual check-in, presenting boarding passes and waiting in line for passport control.

Automation may make travel more accessible to everyone. Process-based changes discussed in this chapter may provide access to tourism for everyone despite a person's ability to drive, their physical condition, age or spoken languages (Cohen & Hopkins, 2019; Ivanov *et al.*, 2017, 2020). As cars become driverless, the requirements of certain physical conditions of the driver are eliminated. This applies to the need for checking the individual's eyesight, avoiding driving when tired or fatigued and following age requirements. Moreover, international travel becomes much easier by removing the language barrier to understanding signs on the road and knowing the driving rules specific to each country. Additionally, the transition between left- and right-steering becomes effortless.

7.4.4 Experience-based innovation

The proposed vision of the future travel experience converts a traveller into a passenger, making the travel part more enjoyable than ever (Pine & Gilmore, 1998). Since the traveller is not preoccupied with the details, schedules, connections or the physical aspect of driving, there is more time to focus on the experience, e.g. taking in the atmosphere, focusing on the new place itself and engaging with people, robots and places. Travellers may appreciate the effortless planning and transit experience, as well as the opportunities to immerse themselves in the destination and connect with significant others. Additionally, tourists may welcome the chance to travel more frequently and explore more remote destinations due to the more affordable cost of travel. Overall, all process- and output-based innovations may give travellers the experience of flow, immersion and efficiency.

On the other hand, not all travellers may perceive changes in the travel process as advantageous and beneficial to them. For example, optimising routes and time of travel is likely to take travellers on the interstate instead of a scenic route that they would have enjoyed more (Cohen & Hopkins, 2019). Other travellers may be disappointed by the reduction in human contact during the travel experience. The entire experience is structured in a way that allows for increased interaction within the travel party; however, random interactions with people en route and at the destination may be limited due to automation and may require more effort to initiate.

7.4.5 Systemic innovation

The evolving transportation landscape will also lead to necessary changes in city planning and the reorganisation of public and private transportation with the core aim of providing a positive experience for everyone. The changes described above are only possible through the increased reliance on system interoperability and data sharing (Buhalis & Leung, 2018; Gretzel, 2022). Data sharing needs to take place across all virtual assistant providers, all vehicles and the overarching algorithm that ties all elements together. The system level should be designed for flexibility (Xiang et al., 2021) to consider modifications for preferences and passenger well-being to take the fastest route or to enjoy the scenic view. In the state of robonomics, a result of all these changes is that the transportation system may encounter a higher burden due to more time available to travel, increased affordability of travel and reduced length en route. Therefore, innovation at the system level is critically important to support all described benefits presented at the process, output and experience levels (Xiang et al., 2021).

7.5 Live

7.5.1 The lodging experience will be customised to 'one'

The growth and development of automation technology will significantly change the way people stay in accommodation (Mercan et al., 2020). Even though the core need to stay in safe, clean lodging establishment will not change (Usta et al., 2011), the way consumers find, book and stay in these properties will change. The focus will move from physical attributes to experience and atmospheric attributes. The process-, output-, experience- and system-based innovations brought by these advancements are presented and discussed below.

7.5.2 Process-based innovation

Studies conducted on lodging establishments for travellers showed very similar attributes as the top factors when consumers select a lodging establishment (Cobanoglu et al., 2003; Pan et al., 2013; Richard & Allaway, 1993; Richard & Sundaram, 1994; Usta et al., 2011). Location, clean room and safety/security often top the list in many studies. These attributes are still a top priority today and will probably be the same in the future. However, what has been changing is how the lodging inventory is distributed and sold, as well as the list of services and amenities offered to guests.

Robotisation and automation will change all processes that guests experience on their stay at a hotel (Buhalis & Leung, 2018). Virtual assistants will play a significant role during the hotel stay (Tung & Law, 2017). The virtual assistant will process check-in at the hotel, connect the guest with the hotel's robots and management and facilitate the review process of the hotel's services and facilities. As guests experience different hotel services (Mercan et al., 2020), they can communicate their likes and dislikes to their virtual assistants. The virtual assistants will channel this information to the hotel, and will also process such information for planning future experiences.

All services at the hotel will be delivered by robots or automated in other ways (Buhalis & Leung, 2018). A robotic associate will recognise the guests as they walk into the hotel, welcome them and help with any questions. All other services of the property will be performed by robotic employees as well, including server, bartender, chef, concierge and housekeeping.

With the implementation of smart surfaces, the process of interacting with the hotel will change. Smart surfaces will provide hotels and hotel guests with the experience of self-cleaning furniture and surfaces, self-dimming windows and touch or proximity surfaces to control the internet of things (IoT) devices and robots at the hotel (Bare Conductive, n.d.).

7.5.3 Output-based innovation

Innovative processes at the hotel will serve as a foundation for increased personalisation and customisation of experiences (Mercan et al., 2020). The overall hotel stay experience will be designed to provide a frictionless journey for guests. Time saving will be realised through uninterrupted transition to the hotel and between the various departments. For example, guests will not have to stop at the front desk to check in, but can complete the check-in process virtually and go straight to their room. Augmented reality (AR) technology may be used for navigation in the guest's phone. AR directions may guide the traveller from their current location to the location of the hotel room saving time and offering convenience (Leung *et al.*, 2020). Virtual assistants can take care of communicating all guest requests to the hotel, freeing up time for the guests to enjoy their vacation. Economically, the hotel experience may become more affordable as robots are introduced on a large scale, and human capital expenses are reduced (Ivanov, 2020). And, finally, virtual assistants, biometric scans and smart surfaces working together may provide guests with ultimate comfort by developing an ideal atmospheric fit to the guest's emotional and physical well-being at every moment of their stay (Ioannou *et al.*, 2020).

7.5.4 Experience-based innovation

The perception and employment of the co-created experience are expected to increase as services become automated (Tussyadiah, 2020). The actual experience of staying at a hotel will extend the benefits offered to customers during the choose and go stages. Customers are likely to evaluate more positive experiences including automatic need recognition, understanding of the customer context, financial and time savings, and highly personalised experiences. While often recognised for assisting with dull, tedious and repetitive tasks (Ivanov *et al.*, 2017, 2020), robots and automation will evolve into providing more suggestion and performing facilitating roles that require deep exploration of available options, knowledge of the tourist's preferences and continuous learning from the lived experiences. The negative side of robotisation is the elimination of humans from the tourist experience (Ivanov, 2020). Therefore, destinations and all hospitality establishments will need to design experiences that will allow guests to interact with local residents, learn about the culture and traditions through human and automated interaction, and partake in some of the activities that showcase the local lifestyle.

7.5.5 Systemic innovation

Innovation at the system level relies on the interconnectedness of all elements of the system that designs, facilitates and provides the tourist experience. Virtual assistants should be able to communicate with hotel management, systems and buildings (Tung & Law, 2017). This will provide guests with the option of full customisation for both the hotel's and the room's physical characteristics such as the colour of the doors and furniture in the room, the art used in the rooms and even the amenities offered to each guest. In addition, hotel rooms will be able to communicate with guests' virtual assistants and identify the desired ambient background colours and music (or white sound) based on the mood of the guest.

At the same time, different levels of permissions should be introduced to protect guest privacy (Ioannou *et al.*, 2020). For example, the scenario at the beginning of the chapter suggested that a guest may choose to activate the entertainment module in the hotel room in order to continue watching the shows that the traveller started watching at home. Similarly, guests may be given an option if they are recognised by facial features, a fingerprint scan or a microchip implant, to use certain hotel services and facilities (Ciftci *et al.*, 2021; Ivanov *et al.*, 2014). Such a system may offer travellers a choice of whether they want to be continuously identified by hotel systems, or whether they would like to use a more on-demand approach (Mercan *et al.*, 2020). While full access to the guest information may offer higher personalisation benefits, it may also come with some perceived intrusion of privacy. Therefore, offering choices may be an effective way to achieve higher satisfaction with the hotel experience.

7.6 Eat

There will be myriad changes, innovations and enhancements to the food service process through automation technologies (Mercan *et al.*, 2020). These changes will influence in-room dining options as well as dining room options, in addition to the various amenities that hotels have traditionally offered to guests. The process innovations that are facilitated by automated technologies will ultimately influence the way in which the guest engages in the dining experience and their expectations. These changes will alter the options available for and the consistency of dining product output as well as the time it takes the food and beverage to be received by the customer. The changes in output will help future hoteliers and restaurateurs satisfy all dimensions of the experience economy, which lead to satisfaction, loyalty and repeat patronage. Finally, the systems used will work seamlessly and efficiently to ensure needs and wants are anticipated, personalised and memorable (Tuncer *et al.*, 2021).

7.6.1 Process-based innovation

The sequence of service process for in-room dining currently involves telephone or online ordering from a limited menu, food preparation and cooking in a designated kitchen and delivery by way of elevator, cart and server to the room (Mercan *et al.*, 2020). The future sequence of service will be streamlined and will eliminate extraneous steps that may lend themselves to service mistakes (e.g. long wait times, cold food, incorrect order) that result in dissatisfaction on the part of the hotel guest (Tuncer *et al.*, 2021). The process will involve infinite choices in terms of cuisine type and recipes, instead of a limited menu, and the choice will be ordered to print in the room on a 3D food printer. Eliminating multiple human touchpoints in the service process diminishes the margin for service errors in addition to food safety errors (e.g. potential contamination points in the food preparation process). Additionally, eliminating between 7 and 10 steps in the process and decreasing wait time while providing more personalised options for each guest reduce room for error and increase the likelihood of customer satisfaction.

The process for a dining room experience becomes similarly more streamlined with the inception of automated services. A traditional sequence of service involves a greet at the host stand, waiting, being brought to a table and handed a menu, waiting, being greeted by a server and having a drink order taken, waiting, being delivered a drink, waiting, having a food order taken, waiting, and then being brought each course with a wait in between, and then a wait for the check (Tung & Law, 2017). The ample waiting time in addition to multiple points of human interaction and transaction leaves room for error.

7.6.2 Output-based innovation

The automated experience will anticipate desires and will already have an inventory of preferences that have been communicated via the personal assistant, regardless of how many times the individual has dined at the given establishment (Tung & Law, 2017). Offerings to which the customer is allergic will not even appear on the virtual menu. Retinal scan from a 3D floating menu coupled with knowledge of food choice preferences quickly filters desired choices including consideration of dietary restrictions and spice-level preferences. Once a choice is made, a perfect pairing of craft cocktails, wine and beer are offered. All cocktails are crafted by robotic bartenders to ensure consistency of quality and speed of production, with the customer notified of the exact wait time before the drink order is served. A physiological composition scan identifies the optimal portion size for the meal in addition to the distribution of protein, vegetables and carbohydrates, as well as the spice levels and the composition of the sauces and garnishes used. Information generated by the scan will be further supported by information relayed by the personal assistant including dietary restrictions, flavour preferences and an assessment of caloric recommendation based on the activity levels exerted that day (Ivanov, 2020).

Once the customer's order is placed, it is instantaneously and 100% accurately uploaded to the 3D food printers, which are used to create the meal within minutes after ordering (Mantihal *et al.*, 2020). Food will be printed from a universal catalogue of recipes that transcend regions, culture and chefs. Options will be limitless. And the time from order to table will be considerably decreased with no preparation or cook times necessary. To set the expectation, as soon as the order is placed, the customer is notified on exactly how long the order will take to create and serve. The food will be delivered immediately on completion by a robot. The food is presented at the perfect temperature in the time provided.

Visually, the food is stunning and the aromas emanating enhance the experience further. The flavours are complimentary based on the preferences of the individual.

At the close of the meal, no physical payment method will be required. The transaction will be deducted from the credit account listed in the personal assistant's database (Ivanov & Webster, 2019). As there will no longer be human staff operating or serving the service environment, a monetary gratuity will no longer exist. However, the robotic servers will collect data about preferences and habits in order to continue to perfect, personalise and streamline the automated service experience based on the information gathered. The result of these outputs will be convenience, time savings and enhanced payment security.

7.6.3 Experience-based innovation

The aim of the output of dining is to capitalise on the experience economy. In terms of dining, in addition to all things travel, tourism and hospitality related, the customer's experience is at the forefront of all actions, accommodations and services provided, which ultimately yield satisfaction, loyalty and repeat patronage. Accordingly, understanding the components of a customer's experience is useful. The output of the automated dining experience facilitates all experiential dimensions established by Pine and Gilmore (1999). For in-room dining, the 3D printer can satisfy the entertainment needs through the 3D printing experience while simultaneously educating the diner about the flavour profiles and ingredients in the dish being prepared. The patron passively participates in the experience while the printing is taking place, and then actively participates in the experience by eating the prepared food. The technology and its product satisfy the aesthetics component through the indulgent environment/product it creates. The automation satisfies the escapism component in that it at once allows the customer to escape *from* their typical dining habits and experiences and escape *to* a hyper-personalised automated dining experience that anticipates desires and needs in the preferred atmosphere and location (Anaya-Sánchez et al., 2019).

The experience allows the guest the option to either have their room transformed into a dining room or physically travel to meet others in an actual physical space in the hotel dedicated to dining. Either way, the ambiance of the room is adjusted to each patron with the dimness or brightness of the lighting, the colour scheme and visual scape, and the type and volume of music all adjusted to the preferences of the customer. The scents emanating from the room are individually and personally crafted to enhance the dining experience without detracting from it.

7.6.4 Systemic innovation

Various systems will work in tandem to create the optimal guest experience. These systems include the personal assistant and the guests themselves, the integrated dining platforms and in-room dining and well-being capabilities, and the integrated hotelscape as an entire smart entity. With the streamlined interconnectivity and communication operating at optimal level, all experiences related to food, beverage and well-being will be tailored to the exact specifications of the individual.

7.7 Exercise

Much like the dining experience of the future, when the customer desires to exercise, an entire automated library of workouts ranging from fitness classes to free weight training to machine work is offered. The room virtually transforms into either a workout studio with all the requisite equipment or transports the customer to a tranquil and restorative location for a zen-like yoga or meditative experience (Yung & Khoo-Lattimore, 2019). The customer will determine the gender and appearance of the trainer for optimal motivation and enjoyment of the experience, as well as the optimal or available workout time.

7.7.1 Process-based innovation

The process for the exercise experience changes considerably in this new paradigm. No longer is there a need to leave one's room and expend time and energy finding the fitness centre. Nor is there a need to wait for the timing of the desired class, which would result in one having to sacrifice experiencing other activities because the set hours for an exercise class conflict with the times of other extracurricular or work-related activities. Instead, the desired fitness programme and trainer may be selected from the catalogue offered through the use of AR and virtual reality (VR), and can bring the exercise world into the hotel room. AR and VR will be supplemented and enhanced through the use of fitness mirrors and in-room equipment to stage an alternate reality experience of exercising in a desirable setting at a time convenient to the guest (Yung & Khoo-Lattimore, 2019).

7.7.2 Output-based innovation

Similar to the restaurant setting, the automated experience will anticipate the desires and abilities of the individual wishing to exercise (Tung & Law, 2017). There will be an inventory of workout preferences that have been communicated from the personal assistant, in addition to health information that will indicate the ability of the individual to partake in a specific workout, the ideal duration and the intensity of the workout based on the physical abilities of the guest. Once a guest chooses the desired workout, they will be provided with the requisite equipment and it will adjust in real time to the ideal cardiovascular and muscular abilities of the individual. Microchip technology can also be used to transmit the physical abilities of the individual and the intensity of the workout can be adjusted as the workout takes place in order to maintain the optimal rate of exertion of the individual.

7.7.3 Experience-based innovation

Through the lens of the experience economy, this new way of working out satisfies the ability to become absorbed in the workout through an immersive experience that caters to the preferences of the individual (Pine & Gilmore, 1999). Additionally, it allows for active and evolving participation in exercise by allowing the guest to leverage AR/VR capabilities and choose the desired type of workout, the ideal environment or setting in which to work out and the desired duration of the workout.

7.7.4 Systemic innovation

Again, the virtual assistant in conjunction with the in-room AR/VR devices will work in tandem to ensure a streamlined and personalised in-room workout experience. Not only will it appeal to the aesthetics of the individual in terms of the VR location of the workout, but it will also adapt on a physiological level to ensure a physical experience that is complementary and beneficial to the individual.

7.8 Review

The virtual assistant may use the post-experience stage to process all feedback it received during the trip. Smart technology that will surround travellers in the state of robonomics may facilitate the long-lasting effects of vacations (Yang *et al.*, 2021), and the review stage may be used as an opportunity for the traveller to review and extend, or even relive the experiences that they had on the trip. For example, a virtual assistant may play a song that the traveller often listened to on vacation, or remind the traveller about a nice meal that they enjoyed during the trip and send the recipe to the 3D printer.

The virtual assistant that has access to the pictures taken at the destinations and the reviews provided by the traveller may be able to generate suggestions for souvenirs, photo books and vacation videos. Souvenirs may be customised with the traveller's photos based on the most memorable experiences. Bringing up memories from trips may be helpful in prolonging the positive impacts of vacations and enhancing the person's well-being. This information will also be used for future vacation planning to further optimise the travel experience (Volchek *et al.*, 2019). Additionally, biometric scans may be used to identify when the next vacation is needed and what type of vacation is needed (Ioannou et al., 2020), thus closing the loop and leading to the beginning of the next cycle of vacation planning.

7.9 Conclusion

This chapter presented a futuristic view of how innovation may shape traveller experiences in the state of robonomics. Employing robots as the main hospitality workforce is predicted to drastically change the processes of how the hospitality product is delivered to the consumer, altering the roles of the organisation, the traveller and technology (e.g. virtual assistant, smart buildings) (Ivanov, 2017; Ivanov et al., 2017, 2020). Such changes will provide various benefits. Customer expectations and satisfaction, in terms of needs and desires, will be met through personalisation, efficiency and accuracy (Volchek et al., 2020, 2021). With properly integrated and communicative systems, the detail of personalisation can satisfy physiological, psychological and hedonic needs. The experiential needs of entertainment, education, aesthetics and escapism may all be satisfied, with emphasis placed on the experience of greatest importance to the customer.

With all the benefits and opportunities that robotisation and automation bring, there are inevitably disadvantages. When the communication between automated devices fails or is not immediate, the threat of dissatisfaction is heightened. When automated services fail, people are far less patient in terms of time for the system to correct itself. Moreover, if details like allergies are assumed to be transmitted, but are not, the ramifications of errors could be life-threatening.

Such processual and output innovations in the service process result in experience-based innovation that encourages travellers to revisit their perceptions of trips. While designed with the best traveller interest at heart, new robotic and automation technologies may sometimes fail and therefore negatively impact the guest experience. Also, one of the key drawbacks of robotisation revealed through the analysis of the different steps of the travel experience is the lack of human interaction (Ivanov et al., 2017, 2020). If needs for human interaction remain unchanged, the hospitality industry will need to find new alternative ways to facilitate interaction between locals and the travelling public.

And, finally, systemic innovation is crucial for supporting the overall guest experience. The state of robonomics will bring major changes to the way people live their lives, not only to the hospitality industry. The changes in the system will be far-reaching, impacting not only the system infrastructure and interoperability in the hospitality sector but also spreading far beyond to cities, countries, policies and laws (Buhalis & Leung, 2018). As data becomes one of the key resources in the state of robonomics, the smooth and efficient transmission of data across different systems becomes essential. In addition, privacy concerns will continue to rise and challenge organisations and governments in defining new approaches and best practices for service customisation and guest protection.

References

Anaya-Sánchez, R., Molinillo, S., Aguilar-Illescas, R. and Liébana-Cabanillas, F. (2019) Improving travellers' trust in restaurant review sites. *Tourism Review* 74 (4), 830–840.

Bare Conductive (n.d.) Smart surfaces will transform the way we live, work and care. See https://www.bareconductive.com/news/smart-surfaces-will-transform-the-way-we-live-work-and-care/ (accessed 3 June 2021).

Buhalis, D. (2019) Technology in tourism-from information communication technologies to eTourism and smart tourism towards ambient intelligence tourism: A perspective article. *Tourism Review* 75, 267–272.

Buhalis, D. and Leung, R. (2018) Smart hospitality: Interconnectivity and interoperability towards an ecosystem. *International Journal of Hospitality Management* 71, 41–50.

Buhalis, D. and Sinarta, Y. (2019) Real-time co-creation and nowness service: Lessons from tourism and hospitality. *Journal of Travel & Tourism Marketing* 36, 563–582.

Buhalis, D. and Volchek, K. (2021) Bridging marketing theory and big data analytics: The taxonomy of marketing attribution. *International Journal of Information Management* 56, 102253.

Ciftci, O., Choi, E.K.C. and Berezina, K. (2021) Let's face it: Are customers ready for facial recognition technology at quick-service restaurants? *International Journal of Hospitality Management* 95, 102941.

Cobanoglu, C., Corbaci, K., Moreo, P.J. and Ekinci, Y. (2003) A comparative study of the importance of hotel selection components by Turkish business travelers. *International Journal of Hospitality & Tourism Administration* 4 (1), 1–22.

Cohen, S.A. and Hopkins, D. (2019) Autonomous vehicles and the future of urban tourism. *Annals of Tourism Research* 74, 33–42.

Globetrender (2016) 2060 forecast: 14 innovations for the hotel of the future. *Globetrender*. See https://globetrender.com/2016/12/14/hotel-of-the-future/ (accessed 3 June 2021).

Gretzel, U. (2022) The smart DMO: A new step in the digital transformation of destination management organizations. *European Journal of Tourism Research* 30, 3002.

Helkkula, A., Kowalkowski, C. and Tronvoll, B. (2018) Archetypes of service innovation: Implications for value cocreation. *Journal of Service Research* 21 (3), 284–301.

Ioannou, A., Tussyadiah, I. and Lu, Y. (2020) Privacy concerns and disclosure of biometric and behavioral data for travel. *International Journal of Information Management* 54, 102122.

Ivanov, S. (2017) Robonomics – principles, benefits, challenges, solutions. *Yearbook of Varna University of Management* 10, 283–293.

Ivanov, S. (2020) The impact of automation on tourism and hospitality jobs. *Information Technology & Tourism* 22 (2), 205–215.

Ivanov, S. and Webster, C. (2017) The robot as a consumer: A research agenda. Proceedings of the 'Marketing: Experience and Perspectives' Conference, 29–30 June, University of Economics-Varna, Bulgaria, pp. 71–79.

Ivanov, S. and Webster, C. (2019) Conceptual framework of the use of robots, artificial intelligence and service automation in travel, tourism, and hospitality companies. In S. Ivanov and C. Webster (eds) *Robots, Artificial Intelligence, and Service Automation in Travel, Tourism and Hospitality* (pp. 7–37). Emerald Publishing.

Ivanov, S., Webster, C. and Mladenovic, A. (2014) The microchipped tourist: Implications for European tourism. In A. Postma, J. Oskam and I. Yeoman (eds) *The Future of European Tourism* (pp. 86–106). Stenden University of Applied Sciences.

Ivanov, S., Webster, C. and Berezina, K. (2017) Adoption of robots and service automation by tourism and hospitality companies. *Revista Turismo & Desenvolvimento* 27/28, 1501–1517.

Ivanov, S., Webster, C. and Berezina, K. (2020) Robotics in tourism and hospitality. In Z. Xiang, M. Fuchs, U. Gretzel and W. Höpken (eds) *Handbook of e-Tourism* (pp. 1873–1899). Springer. https://doi.org/10.1007/978-3-030-05324-6_112-1

Leung, X.Y., Lyu, J. and Bai, B. (2020) A fad or the future? Examining the effectiveness of virtual reality advertising in the hotel industry. *International Journal of Hospitality Management* 88, 102391.

Mantihal, S., Kobun, R. and Lee, B.B. (2020) 3D food printing as the new way of preparing food: A review. *International Journal of Gastronomy and Food Science* 100260.

McKercher, B. (2016) Towards a taxonomy of tourism products. *Tourism Management* 54, 196–208.

Mehrabian, A. and Russell, J.A. (1974) The basic emotional impact of environments. *Perceptual and Motor Skills* 38 (1), 283–301.

Mercan, S., Cain, L., Akkaya, K., Cebe, M., Uluagac, S., Alonso, M. and Cobanoglu, C. (2020) Improving the service industry with hyper-connectivity: IoT in hospitality. *International Journal of Contemporary Hospitality Management* 33, 243–262.

Moyle, B.D., Moyle, C.-L., Bec, A. and Scott, N. (2019) The next frontier in tourism emotion research. *Current Issues in Tourism* 22, 1393–1399.

Negri, N.A.R., Borille, G.M.R. and Falcão, V.A. (2019) Acceptance of biometric technology in airport check-in. *Journal of Air Transport Management* 81, 101720.

Neuhofer, B., Buhalis, D. and Ladkin, A. (2014) A typology of technology-enhanced tourism experiences. *International Journal of Tourism Research* 16, 340–350.

Pan, B., Zhang, L. and Law, R. (2013) The complex matter of online hotel choice. *Cornell Hospitality Quarterly* 54 (1), 74–78.

Pearce, P.L. (2011) *Tourist Behaviour and the Contemporary World*. Channel View Publications.

Pine, B.J. II and Gilmore, H.J. (1999) *The Experience Economy: Work is Theatre & Every Business a Stage*. Harvard Business School Press.

Prentice, C. (2019) Emotional intelligence and tourist experience: A perspective article. *Tourism Review* 75 (1), 52–55.

Prideaux, B. (2019) Drive and car tourism: A perspective article. *Tourism Review* 75 (1), 109–112.

Ramaswamy, V. and Ozcan, K. (2018) What is co-creation? An interactional creation framework and its implications for value creation. *Journal of Business Research* 84, 196–205.

Richard, M.D. and Allaway, A.W. (1993) Lodging choice intentions: A causal modeling approach. *Journal of Hospitality & Leisure Marketing* 1 (4), 81–98.

Richard, M.D. and Sundaram, D.S. (1994) A model of lodging repeat choice intentions. *Annals of Tourism Research* 21 (4), 745–755.

Shin, H., Perdue, R.R. and Kang, J. (2019) Front desk technology innovation in hotels: A managerial perspective. *Tourism Management* 74, 310–318.

Tuncer, I., Unusan, C. and Cobanoglu, C. (2021) Service quality, perceived value and customer satisfaction on behavioral intention in restaurants: An integrated structural model. *Journal of Quality Assurance in Hospitality & Tourism* 22 (4), 447–475.

Tung, V.W.S. and Law, R. (2017) The potential for tourism and hospitality experience research in human–robot interactions. *International Journal of Contemporary Hospitality Management* 29, 2498–2513.

Tussyadiah, I. (2020) A review of research into automation in tourism: Launching the *Annals of Tourism Research* curated collection on artificial intelligence and robotics in tourism. *Annals of Tourism Research* 81, 102883.

Tussyadiah, I.P., Zach, F.J. and Wang, J. (2017) Attitudes toward autonomous on demand mobility system: The case of self-driving taxi. In R. Schegg and B. Stangl (eds) *Information and Communication Technologies in Tourism 2017* (pp. 755–766). Springer.

Usta, M., Berezina, K. and Cobanoglu, C. (2011) The impact of hotel attributes' satisfaction on overall guest satisfaction. *Journal of Service Management* 6 (3), 1–12.

Vargo, S.L., Akaka, M.A. and Wieland, H. (2020b) Rethinking the process of diffusion in innovation: A service-ecosystems and institutional perspective. *Journal of Business Research* 116, 526–534.

Vargo, S.L., Lusch, R.F., Akaka, M.A. and He, Y. (2020a) Service-dominant logic. In E. Bridges and K. Fowler (eds) *The Routledge Handbook of Service Research Insights and Ideas* (pp. 3–23). Routledge.

Volchek, K., Law, R., Buhalis, D. and Song, H. (2019) The good, the bad, and the ugly: Tourist perceptions on interactions with personalised content. *e-Review of Tourism Research* 16 (2/3), 62–72.

Volchek, K., Law, R., Buhalis, D. and Song, H. (2020) Exploring ways to improve personalisation: The influence of tourist context on service perception. *e-Review of Tourism Research* 17, 737–752.

Volchek, K., Yu, J., Neuhofer, B., Egger, R. and Rainoldi, M. (2021) Co-creating personalised experiences in the context of the personalisation-privacy paradox. In *Information and Communication Technologies in Tourism 2021: Proceedings of the ENTER 2021 eTourism Conference, January 19–22, 2021* (pp. 95–108). Springer International Publishing.

Yang, H., Song, H., Cheung, C. and Guan, J. (2021) How to enhance hotel guests' acceptance and experience of smart hotel technology: An examination of visiting intentions. *International Journal of Hospitality Management* 97, 103000.

Yanishevskaya, N., Kuznetsova, L., Zhigalov, A., Parfenov, D. and Bolodurina, I. (2019) Development of an intellectual module for selection of places to travel in the virtual assistant system for planning trips. *Journal of Physics: Conference Series* 1399 (3), 033059.

Yung, R. and Khoo-Lattimore, C. (2019) New realities: A systematic literature review on virtual reality and augmented reality in tourism research. *Current Issues in Tourism* 22 (17), 2056–2081.

Xiang, Z., Stienmetz, J. and Fesenmaier, D.R. (2021) Smart tourism design: Launching the *Annals of Tourism Research* curated collection on designing tourism places. *Annals of Tourism Research* 86, 103154.

8 Development of Robot-Friendly Hospitality Facilities

Katerina Berezina, Olena Ciftci and Fernando Arroyo Lopez

8.1 Introduction

Robotics and automation have brought significant changes to the hospitality and tourism industry and will continue to further revolutionise the sector as it moves to the state of robonomics (Ivanov et al., 2017, 2020; Webster & Ivanov, 2020). The new economic state of robonomics suggests that human employees are being replaced by robots, artificial intelligence and automation (Webster & Ivanov, 2020). As robots form the core workforce of the hospitality industry, the facilities will need to evolve to provide a friendly environment to the new type of employees (Ivanov & Webster, 2017b). The required changes will go beyond adjusting the existing hospitality establishments that employ a human-centred design and will call for a radical rethink of the facility design.

This chapter offers a discussion of the changes that may lie ahead for the hospitality industry as it transitions to robonomics and modifies hotels, restaurants, event venues and other facilities to meet the needs and demands of robotic employees. As robots and automation penetrate the industry, they will change the way people travel, create new business models and eliminate some existing businesses, and will modify processes and standard operating procedures in the hospitality industry. All tourism and hospitality sectors, including transportation and travel, hotels and restaurants, will be affected. The discussion that follows presents potential scenarios of how the state of robonomics may affect the tourism and hospitality industry. It is important to envision the potential changes that may take place in the industry because new types of vehicles or businesses will require new unique designs, and modifications to service processes may alter the layouts of hospitality facilities.

Changes in the *transportation* sector will alter the way people travel, potentially blurring the lines between transportation and accommodation. As self-driving vehicles become mainstream, people will be able to travel longer distances by car more safely and cost-effectively than ever before (Cohen & Hopkins, 2019). Additionally, the travel experience will evolve because tourists will no longer have to focus on the driving task. Instead, travellers will be able to enjoy time with their family and friends, sightseeing, relaxing, eating, having sex or engaging in other activities as the car drives. Therefore, the automobile may become more of a 'social space' than just a vehicle. Such changes in the transportation sector may call for necessary changes in the hotel industry, leading to the creation of a new hotel type: plug-and-play hotels.

Plug-and-play *hotels* are the new hotel concept that consists of a hotel stem where self-driving hotel rooms (SDHR) (see Photo 8.1) plug in to offer tourists the comfort of spacious hotel rooms, as well as access to a larger hotel infrastructure and services. This concept may be viewed as analogous to the current recreation vehicle (RV) site on steroids. However, instead of parking on an assigned parking spot, the SDHRs will be able to attach to the hotel building, the stem that carries the core hotel infrastructure and expandable hotel rooms. Once properly sealed, the connector doors will unlock, opening access to the hotel room and from there to the rest of the facility. The design of SDHRs needs to be strategically aligned with the design of the hotel stem, which is likely to lead to partnerships between self-driving vehicle manufacturers and hotel brands. While the same SDHRs may be used by all travellers, the new hotel concept would allow for brand and service-level differentiation by creating different stem building designs. For example, an economy hotel may offer a smaller room, breakfast in the lobby lounge, a pool and housekeeping and laundry services. However, an upscale hotel may offer larger rooms with dedicated living, dining and working areas, and a balcony, as well as a larger hotel infrastructure inclusive of pools, spas, a gym, several restaurants and a variety of services, such as room service and hairdressers.

Traditional hotels as we know them today will coexist with plug-and-play properties by serving another customer segment travelling by airplanes, trains and cruises. Some of these hotels may be newly built with robotic employees in mind and specifically designed for that purpose. However, some of the properties may be renovated to meet the needs of the new workforce. For example, historical hotels may be protected from demolition or extensive renovation but instead will undergo necessary adjustments to make the work environment robot-friendly. Such adjustments may include changes to the exterior and interior design, rethinking space use and internal hotel processes, as discussed further in the chapter.

Changes in the transportation sector and travel patterns will impact not only hotels but also other hospitality sectors. As self-driving vehicles will free up the driver and make it more comfortable to eat on the road, more *restaurants* may need to offer take-out services and redesign their facilities for a smooth to-go pick-up process. On the other hand, restaurant concepts focused on dining in the car (e.g. Sonic, an American drive-in quick service restaurant) may become obsolete because travellers would not have to stay parked next to the restaurant building while consuming their meal. Therefore, such restaurants will need to be redesigned or repurposed.

Additionally, hospitality businesses may no longer need *parking* garages on their property. Currently, a large portion of real estate is allocated to parking lots that would host the cars of all hotel guests, restaurant patrons or convention centre visitors. As self-driving or autonomous, vehicles make their way into societies, a city's infrastructure will need to be adjusted to become smarter and more efficient (Cohen & Hopkins, 2019; Computing Community Consortium, 2020; Stone *et al.*, 2016). Parking garages may be located in different areas of the city to serve the needs of residents and travellers. However, they no longer have to be located right next to the building as self-driving vehicles will be able to drop off passengers by the front door of the business they are visiting, self-park and then pick up the passengers when needed. Such changes fall into the city planning domain; however, from the perspective of the hospitality industry, they signal an opportunity to redesign vast parking spaces into other facilities offering meaningful experiences to guests or abandon the space to save money.

In addition to these big-picture changes, the arrival of robots and the transition to robonomics will call the hospitality industry to rethink its processes and standard operating procedures, which in turn, may influence the design of the facility. The type of robots used, e.g. stationary vs mobile robots, will drive decisions about the design and use of space in hospitality facilities. While stationary robots need an organised workplace conducive to their responsibilities and tasks, mobile robots impose more requirements on the design of the building to ensure their easy navigation. As the state of robonomics assumes that hospitality facilities are entirely staffed by robots, mobile robotic employees will need access to all areas of the hospitality facilities.

In summary, designing robot-friendly facilities is a complex process that this chapter unfolds. The chapter progresses by examining the definition of robot-friendly spaces. From the definition, the chapter moves to consider the human–robot perspective on design. Even though this chapter focuses on creating robot-friendly environments, it is important to keep in mind that hospitality facilities need to be designed in such a way as to accommodate human and robotic customers and employees. Keeping the human–robot perspective on design as an overarching theme, the chapter further presents three perspectives on designing robot-friendly hospitality facilities: (1) impacts on hospitality operations, (2) system design and (3) architectural perspectives. Figure 8.1 presents the framework of a robot-friendly facility design built based on the discussions presented in this chapter.

Photo 8.1 Self-driving hotel room (Generated by the authors using Image Creator in Microsoft Bing)

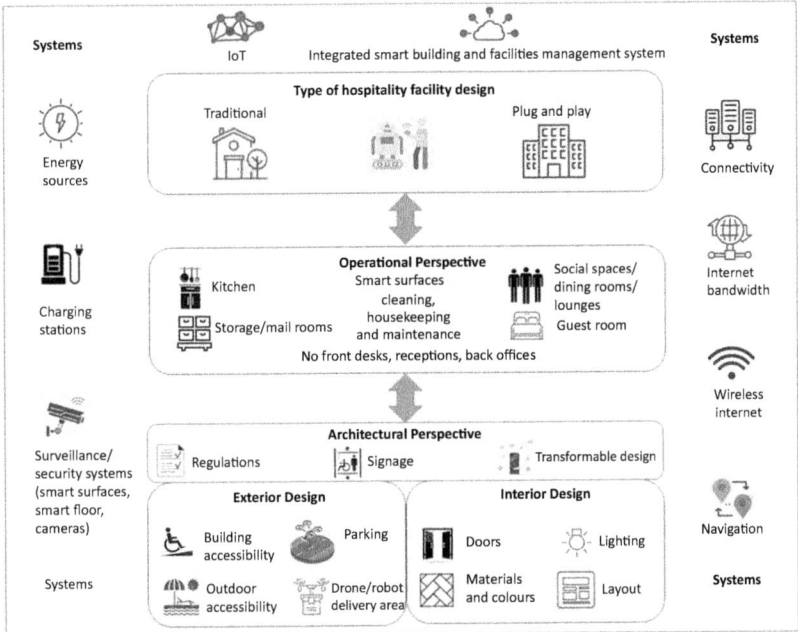

Figure 8.1 The framework of robot-friendly facility design

8.2 Robot-Friendly Design

Robot-friendly facility design may be defined as the type of design that is planned and organised with the robot, its abilities and tasks in mind (Ivanov & Webster, 2017b; Tan *et al.*, 2016). The robot-friendly design supports the robot's ability to navigate through the environment, identify and avoid obstacles, reach for necessary objects and interact with other robots and humans while protecting the safety of all occupants of such space. While robotic technology is developing every day, it is still challenging for a robot to go up a stairs, turn a door handle or knock on a door. These actions, which may not seem like major demands for a human body, may significantly interfere with robots' activities. Therefore, it is imperative to plan hospitality facilities in a way that would eliminate such hindrances and create robot-friendly environments. Additionally, robots have some needs that human beings do not, e.g. using a charging station or downloading a digital facilities map. As humans, we also have a need to recharge and learn about our environment. However, we usually satisfy these needs differently (e.g. by consuming food, sleeping or reading property signs). When creating robot-friendly facilities, it is important to expand our thinking and include robots in our consideration of factors necessary for creating comfortable, convenient and safe spaces. The robot-friendly facility design has also been referred to as a robot-inclusive design (Tan *et al.*, 2016).

Robot-friendly facility design is a critical component of the success of hospitality businesses operating in the state of robonomics. First, the robot-friendly design supports robots' ability to perform their intended tasks, which, in the state of robonomics, means all tasks around the property, from checking in a guest to cleaning the facility, cooking and serving food, and staffing a conference reception. As robots become the main workforce in the hospitality industry, their performance will determine the failure or success of the business. Second, a robot-friendly facility may make less sophisticated robots successful in that environment. Investing in a robot-friendly facility design may contribute to the company's bottom line, as less sophisticated robots are usually less expensive. Third, robot-friendly facilities are usually 'smart' facilities that not only support the work of robots but also create an infrastructure for enhancing guest experiences. Finally, robots enable the effectiveness and efficiency of the industry; therefore, a friendly design will help with customer satisfaction and the bottom line.

8.3 Human–Robot Perspective on Design

As the hospitality and tourism industry transitions to the state of robonomics, it is likely that it will employ and serve both humans and robots (Ivanov & Webster, 2017a). Therefore, the design of robot-friendly hospitality facilities faces the challenge of creating an environment that is supportive of the needs of robotic employees and, at the same time, is relaxing, memorable and visually appealing to human guests. Several studies have found that the physical environment of any service setting has an impact on the emotional response of an individual, which later evokes either a positive or a negative behaviour towards the named environment (Countryman & Jang, 2006; Kim & Moon, 2009; Lockwood & Pyun, 2019). The physical environment has also been proven to affect consumer expectations (Hanaysha, 2016; Simpeh *et al.*, 2011), perceived value and image (Ali *et al.*, 2013; Han & Hyun, 2017), hotel booking intentions (Nanu *et al.*, 2020), consumer satisfaction (Ryu *et al.*, 2012; Tutuncu, 2017), delight (Ali *et al.*, 2016b) and loyalty (Alexandris *et al.*, 2006; Ali *et al.*, 2016a). Therefore, while supporting robots, robot-friendly hospitality facilities should be designed to drive favourable customer responses.

While the main purpose of hospitality facility design is guests' comfort and a pleasant and memorable experience, at the same time, the design should be functional, support navigation and enable the efficient completion of tasks by robotic employees (i.e. robot-friendly). With the development of technologies, the design of hospitality facilities will include transformable elements that will change according to customer needs, e.g. the location and shape of the furniture may be changed and walls may be removed or assembled in a very short time (Cao, 2020; see Section 8.6.3). Changes in the design should not create obstacles for robots' navigation, completion of tasks, providing service and ensuring the safety and security of the guests. The following sections discuss the essential combination of hospitality facilities' design elements, systems and rules that ensure a robot friendly and, at the same time, a comfortable environment for humans (guests and employees).

8.4 Operations Perspective

From the operations perspective, when entering the state of robonomics, the hospitality industry may need to adjust its operating procedures, which may lead to changes in the facility design. Such changes may impact various departments of the hospitality businesses, including the front office, kitchens, guest rooms, social spaces (e.g. lobbies, pools and dining rooms) and meeting spaces. The main considerations that should be evaluated by the hospitality industry include the purpose and use of the space, accessibility to the space for cleaning and navigation purposes, and ease of access to different objects needed for business operations.

8.4.1 Front of house

With robotic employees on board, the lobby areas of hospitality facilities may evolve into more social lounge-type spaces. In the hotel sector of the industry, the front desk is likely to disappear as robotic employees may be equipped with all the necessary technologies (e.g. biometric and document scans and payment processing) to complete guest check-ins at any location. For that reason, there will be no need to attach the check-in function to a specific location, such as a front desk. Instead, customers may be checked into the hotel while relaxing on a couch in the lobby. Such a change in hotel operations may lead to a design change as the entire lobby area may be transformed into a lounge, an area with comfortable seating and working areas surrounded by dining and shopping outlets. We may see similar changes at convention facilities. As check-in services may be automated, more room in the lobby area will be left for networking and socialising.

8.4.2 Back of house

As hospitality facilities become more automated, the back of house may need to be reorganised. Some functions (e.g. human resources) may become irrelevant in the state of robonomics. At the same time, some tasks in other areas (e.g. marketing, accounting, revenue management) may be automated to a certain extent, centralised and supervised by humans remotely. Such operational changes will require hospitality facilities to be designed in such a way as to host both human and robotic employees and provide an environment convenient and comfortable for both. Additionally, employee lounges will need to be redesigned to fit the needs of robotic employees. Changes to the back-of-house office space call for revisiting the property layout. It may also free up some space that may be used for guest services.

8.4.3 Kitchen facilities

When considering the robot-friendly design of hospitality facilities, kitchen areas deserve special attention. Kitchen spaces are usually crowded with people and equipment. Such environments introduce challenges for robotic employees to access and operate in them. The state of robonomics will require revisiting the kitchen design to plan the area conducive to robots performing their intended functions. Robotic chefs are usually stationary robots and may be designed as full-body robots, robotic arms, robotic sushi-makers or other types of robots specialising in preparing certain dishes (Berezina *et al.*, 2019). Stationary robots do not present many challenges with design. However, it is important to strategically place robotic chefs and plan the rest of the kitchen operations around them. For example, robotic chefs will need access to the food storage areas and freezers, or delivery robots will need to restock the supply of ingredients used in cooking. Additionally, robotic waiters will need to pick up plated dishes for delivery. Therefore, kitchen facilities should be planned and laid out to support navigation to and within the kitchen.

8.4.4 Cleaning, housekeeping and maintenance

In the future, with the development of smart surfaces, all surfaces in hospitality facilities will be self-cleaning and coated with a hydrophobic and antibacterial coating (Ali & Ali, 2020; Bare Conductive, 2020). The smart surfaces will not require regular cleaning and sanitation (Ali & Ali, 2020; Atwa et al., 2015; Bare Conductive, 2020). Curtains, bed/bath linen and tablecloths will be made from long-lasting fabrics that repel dirt and also require less cleaning and water to clean them. Walls and pavements made of damage-resistant and self-healing materials will not require routine maintenance (Ali & Ali, 2020; Atwa et al., 2015; Bare Conductive, 2020). Also, smart surfaces will notify the integrated smart building and facilities management system (ISBFMS) (Hotel Innovation Committee, 2019) when it is time for cleaning or maintenance (Bare Conductive, 2020). The ISBFMS will assign robots to perform housekeeping or maintenance tasks. The smart surfaces will monitor cleaning and maintenance events and report them to the ISBFMS (Bare Conductive, 2020).

Robot-friendly spaces should be designed with consideration of how spaces will be cleaned. Observability, accessibility, activity and safety are the four exemplary design principles to successfully deploy a floor-cleaning robot (Elara et al., 2014). Cleaning performance measured by floor coverage was found to be 21% higher in a robot-inclusive space as compared to a non-robot-friendly space (Elara et al., 2014). Shape is one of the important elements to consider when designing robot-accessible spaces. Rectangular and circular forms are easier to clean with robotic cleaners, whereas edged angles and irregular forms may leave some spaces unreachable for robots (Ivanov & Webster, 2017b).

The design and quantity of furniture should also be considered. Hotel guest rooms as well as various social spaces, e.g. dining rooms, lounges and banquet and conference rooms, usually have a lot of furniture that robots need to navigate around to clean those spaces. In plug-and-play hotels, robots will also need to clean the self-driving rooms when they are attached to the hotel stem. These rooms are sized like RVs and are packed with furniture to provide comfort to travellers while on the road. Therefore, the design of these spaces becomes even more important.

In addition to planning the space, it is important to provide robots with access to the necessary cleaning supplies. All housekeeping and storage rooms (e.g. linen closets) should be laid out for easy access. Smart technology may be used to assist with some housekeeping functions, such as restocking and refilling. Smart internet of things (IoT) sensors on soap and toilet paper dispensers will send notifications so that a robot can be sent to fill these to optimal levels established by the management system.

8.5 System Design

With the increased automation and robotisation of hospitality facilities, the industry will greatly rely on connectivity and networks. Therefore, when designing a hospitality facility, it is important to consider the information and communication technology (ICT) infrastructure and ensure that all hospitality facilities are properly 'wired' to accommodate the needs of the robotic employees. The areas that require special attention include device connectivity, e.g. robots, sensors, IoT devices, as well as navigation, power sources and surveillance systems.

8.5.1 Connectivity

Connectivity and networking are some of the crucial steps in ensuring the success of robot-staffed hospitality facilities. As peer-to-peer communication across all devices may be expensive, communication from each device to a central controller, e.g. a facility mastermind, may be recommended (Mazzara *et al.*, 2019). The ISBFMS may serve as such a central controller for hospitality facilities. The internet bandwidth should be carefully evaluated to support robotic employees, work processes, guest connectivity needs and guest robotic assistants. While stationary robots may have a wired connection, mobile robots will rely on wireless networks. Therefore, both types of networks should be thought through and designed with appropriate cabling and access points.

8.5.2 Navigation

The navigation of mobile robots needs to be supported by a digital map of facilities (Berezina, 2018; Ivanov & Webster, 2017b). Such a map may be pre-loaded on all robotic employees of the facility. At the same time, the map should be accessible via communication with the central controller and dynamically updated. Given the opportunities brought about by transformable technologies, the layout of the hospitality facilities may be changed regularly (see Section 8.6.3). Therefore, there is a need for all robotic employees to be up to date on the latest facilities map. Additionally, hospitality facilities may offer new services in the domain of robot navigation, such as access to the dynamic digital property map by a family robot that travels with guests or a robotic assistant that accompanies a business traveller at a conference.

Dynamic facility maps will be updated based on the data from smart floors. Smart floors can detect the location and type of objects (e.g. furniture, people, other robots) in the area via sensors (Bourée & D'Andrès, 2018). The data from smart floors will be sent to the ISBFMS. The ISBFMS will process the data and update dynamic facility maps in real time. Dynamic maps will also include the locations of electronic markers visible to robots and invisible to humans. Robots will build their path inside and outside the property using dynamic digital property maps and electronic markers. While creating the navigation path for themselves, robotic employees will follow the navigation priority rules. They will yield right of way to human guests first with guest robots given second priority and the third right of way will be given to human employees.

8.5.3 Energy sources

The robot-friendly hospitality facility design should take into account the issue of how companies' and guests' robots will be powered and recharged. As 'the charger continues to be the most common destination' for robots (Pinillos *et al.*, 2016), hospitality facilities should provide plenty of charging stations for both robotic employees and robotic assistants brought by guests. The hospitality and tourism industry had already experienced a power outlet shortage when guests started bringing more devices on their trips (Berezina, 2018). Similarly, airport waiting areas had to be equipped with charging stations to support the needs of the travelling public. In the state of robonomics, hospitality facilities, including hotel rooms and conference venues, will have to incorporate facilities that enable guests to recharge their robots (Ivanov & Webster, 2017b).

While the energy demands of robot-staffed hospitality facilities will increase, the smart building infrastructure and ICT may assist in balancing the heightened power needs. For example, the ISBFMS will help to reduce energy and water consumption, the environmental impact and operating costs. The IoT technologies will monitor data from different equipment (e.g. chiller energy data, ambient temperatures, food temperature and production, waste, water quality and inventory levels) to ensure optimal conditions for the work of the equipment, minimise waste and maximise profits. These measurements and actions will result in the prediction of maintenance, where attention will be given to the prevention of potential problems. Using the smart water management system provides hotels with an opportunity to configure the water flow in such a way that would lead to reducing costs while maintaining a positive guest experience (Smart Hotel Technology Guide 2019, 2019).

8.5.4 Surveillance systems

Surveillance systems will include smart surfaces and smart floors that can monitor occupancy and issues of safety and security (e.g. ability to detect if somebody falls on the floor) (Bare Conductive, 2020; Bourée & D'Andrès, 2018). Smart surfaces, smart floors and surveillance cameras should communicate with the central controller of the facility. If an emergency or threat is detected, the central controller will dispatch security robots based on their proximity to the location. The controller would also initiate transformable technologies (see Section 8.6.3) if needed, for example, to create additional ramps for robots to speed up their arrival at the site. At the system level, surveillance cameras, smart floors, digital facility maps and transformable technologies should all work together to guarantee safety and speed up the arrival of security robots to the location where they are needed.

8.6 Architectural Perspective

As robots will assist or perform various hospitality tasks, they will work inside and outside facilities (Ivanov & Webster, 2017b; Webster & Ivanov, 2020). Thus, the design of both the interior and exterior spaces of hospitality establishments should be robot friendly. While hospitality companies fully determine and control the design of the interior spaces of their establishments, the exterior design and, thus, its robot-friendliness may not be under the control of the company (Ivanov & Webster, 2017b). For example, when a restaurant, hotel or spa is located within other businesses, such as shopping malls or airports, the design of the parent facility will influence the design of the hospitality facilities nested in it. We envision that service robots will work in all kinds of service facilities, including airports and shopping malls; thus, they should also adjust or build their indoor and outdoor spaces to be robot friendly.

One example of a facility-wide, robot-friendly adjustment is the elimination of doorsteps from the interior and exterior design and instead use ramps to accommodate robots. This design will allow mobile service robots to better navigate through the facility, optimise mobilisation and allow cleaning robots to perform their task better. It will also help hospitality facilities to provide access to people with disabilities.

8.6.1 Impacts of government regulations on design

The changes proposed in this chapter will be moderated by the regulations that emerge as society transitions to the state of robonomics. The interior and exterior design of robot-friendly facilities should follow local, state and federal regulations (Ivanov & Webster, 2017b). Main attention should be given to hallway width, fire regulations, safety regulations and capacity restrictions accounting for both humans and robots. The hallways and corridors need to be wide enough to allow the movement of people and robots at the same time and follow safety regulations.

The hospitality facilities, their outdoor areas, recreation facilities, swimming pools, sport and boating facilities, golf courses and amusement rides are already designed as accessible facilities that account for the needs of people with disabilities (United States Access Board, n.d. b). In the United States, these facilities should be built in accordance with the Americans with Disabilities Act (ADA) and the Architectural Barriers Act (ABA) (United States Access Board, n.d. b). In robonomics time, these regulations will be modified to accommodate people with disabilities and their service robots. For example, the new regulations on public rights of way (United States Access Board, n.d. a), in the future, may require people and robots not to obstruct the movements of people with disabilities or their robotic assistants. Thus, following the new regulations, the design of outdoor and indoor spaces will be adjusted to allow both a person with disabilities and a service robot to move around in the same area and access desired areas at the same time without interfering with each other.

8.6.2 Signage

Even though robots will navigate facilities using the digital map of facilities (Ivanov & Webster, 2017b), they will interact with humans and other objects that will be detected by their cameras. Thus, using artificial intelligence, robots should be able to properly respond to signs. Hospitality facilities will need to accommodate special guidelines on the use of signage, considering that robots and humans will interact at the same time in the same place. The guidelines will provide humans and robots with a proper way to communicate and convey information that will help them in their decision-making based on the information given by the sign. For example, the signs will provide information about the limits on room capacity, allowed weights, movements, circulation and what to do in case of an emergency. As an example, elevator capacity and weight limits will be required for both humans and robots. In addition, the signs need to follow national, state and local regulations.

8.6.3 Transformable design

Transformable interior and exterior design will become common in the future (Cao, 2020). Hospitality facilities will have retractable walls and storage, transformable furniture and multi-purpose shelving (Cao, 2020). The transformable design will create multi-functional spaces (Cao, 2020) and also allow effective and efficient robot cleaning of the spaces. Also, part of the building, wall or fence may transform into an open area to allow access for autonomous trucks or delivery robots to the facility if the space around the facility is limited. Steps can also be transformed into ramps or moving platforms to accommodate the mobility of service robots. The transformable elements of design will be integrated with smart building technology and will be controlled by robots and the ISBFMS.

The development of nanotechnology will allow hospitality facilities or their parts to morph and self-assemble (Ali & Ali, 2020; Canton, n.d.). Such transformations will present the opportunity to modify some parts of plug-and-play properties to accommodate different types of SDHRs and optimise hotel capacity. The design of a building facade, including its colours and images, can also be changed, for example, based on the theme of a big event or customer preferences. Furthermore, the in-room design may also be adjusted based on guest preferences.

8.6.4 Exterior design

In the state of robonomics, city spaces and outdoor spaces will be transformed to accommodate service robots and autonomous vehicles (Cohen & Hopkins, 2019; Computing Community Consortium, 2020). For better outdoor mobility, all structures that are designed with accessibility in mind, including wheelchair ramps and automatic doors, can be used by service robots that move on wheels (Ivanov & Webster, 2017b). Thus, designing accessible buildings and spaces is one of the major steps in the direction of creating robot-friendly facilities.

Robots and drones already deliver food (Berezina *et al.*, 2019) and packages (Palmer, 2020; Ward, 2020). With the development of technology, the automated delivery of freight, packages and food will prevail (Computing Community Consortium, 2020; Stone *et al.*, 2016). Delivery robots may be owned by the hospitality company (e.g. restaurants) or a third-party delivery company (e.g. Amazon, Walmart, Uber Eats contractors). Thus, delivery robots and drones, as well as autonomous trucks, need to have a designated parking area for loading/unloading packages or food accessible by other service robots that will load/unload and take the packages inside a facility.

The guests will arrive at hospitality establishments on autonomous driving or flying vehicles (e.g. passenger drones). While vehicles such as cars or capsules can be parked underground or use remote parking, passenger drones need open spaces such as ground or roof parking. Autonomous passenger drones will be similar to manned passenger drones that are already operating and carrying people (DroneTrader, 2020). Passenger drones are small compared to helicopters (the size of a car) and take off and land vertically; thus, they do not need a big area for take-off and landing or runways (EHang AAV, n.d.). Autonomous vehicles will be electric powered. Therefore, charging stations at parking lots (if not prohibited by regulations or law at that time) may provide an additional service for the guests.

8.6.5 Interior design

To design robot-friendly spaces inside a hospitality facility, architects and designers need to consider changes or adaptations to the doors, materials, colours, lighting, layout, floor surface, signage within the building, the width of the corridors, stairs, lifts, artificial and natural landmarks, among other elements.

8.6.5.1 Doors

Because robot-friendly facilities will house mobile service robots, doors will be one of the main changes to their interior design. Traditional hotels customarily use side-hung doors that open with door handles for most of their spaces, but to facilitate and optimise navigation for robots, the implementation of automated sliding doors will be imperative. Robots will rely on smart IoT sensors, the digital map of the facility and the ISBFMS that will exchange information with the automated doors, allowing robots to enter the space behind the door without any interruption or identification. For example, autonomous luggage delivery carts will bring suitcases from a hotel's entrance to the rooms without any identification or stops required by a door.

Doors will be connected with smart cameras with facial recognition capabilities through the ISBFMS based on the photo submitted during mobile check-in. This will give customers access to the facilities, reduce contact points with public surfaces and minimise the time spent on authentication when moving around the facility. IoT sensors will track real-time room occupancy so that robots will not enter a room that is being occupied. For example, a restaurant guest will enter a restroom facility through an automatically opened door that will automatically lock and robots will not enter the room until the guest exits the room without being disturbed.

8.6.5.2 Materials and colours

The colours of furniture, walls, floors and other surfaces should comply with the design principles that appeal to humans since they will be the main customers of the hospitality facilities. According to Treptow *et al.* (2004), as the computer vision field continues to improve, visual cues that are used nowadays in some robots will be removed in the near future. These visual cues may be replaced with electronic markers visible to robots only but invisible to humans. Therefore, it is predicted that colours used in the interior design will not affect the robots' ability to navigate through the space and function.

When architects and designers select the materials for floor surfaces and furniture, they need to consider the fact that they will be cleaned by a robot. Therefore, selecting materials that will not be easily detached by the cleaning robot's movement is imperative. This same guideline will apply to service robots involved in cleaning restrooms, pools, gyms, windows, etc. Avoiding damage to the robot's mechanisms would also be taken into consideration.

As previously stated, when designing robot-friendly facilities, it is important to keep human preferences in mind as they will remain the main customers of hospitality establishments. From this perspective, robots employed by the hospitality industry contribute to the feel and look of an establishment. Additionally, the friendly demeanor of the guests may become one of the aspects of robot-inclusive environments. To increase interactive efficiency between humans and robots, the use of colour can play an important role in provoking emotions in people. Song and Yamada (2017) suggested using the colour white on robots that have no facial expressions to embody a relaxed emotion in the observer. In their study, Löffler *et al.* (2018) suggested that joy is understood in terms of warmth and light and, thus, can be expressed through the colour yellow and high-pitched tones, or a combination thereof.

8.6.5.3 Lighting

All bulbs and lamps will be connected through IoT sensors to a lighting management system (LMS) as a fundamental part of the ISBFMS (Hotel Innovation Committee, 2019). Lighting will be adjusted automatically in rooms and public spaces depending on guest preferences, the amount of ambient light required, or predetermined levels by the management. Robots will not require any lighting as they will navigate through the digital map of facilities and have a built-in camera with night vision that will allow them to detect objects or persons on their way. This ability of robots will contribute to efforts in reducing energy consumption. The LMS will alert the maintenance team if a bulb needs to be changed so that a maintenance robot will fix it.

8.6.5.4 Layout

Robot-inclusive spaces will need changes to some architectural and design features to optimise the performance of service robots. Mobile service robots need to easily access and navigate a floor area by arranging furniture along the walls, combining the legs of different pieces of furniture, elevating furniture to make more available space under it and creating large open areas. The main goal in designing the layout of a robot-friendly facility will be to optimise the flow of movement and the circulation of humans and robots at the same time. Therefore, changes in the size of public and private spaces will be required, as well as the form of those spaces and the arrangement of furniture and objects.

8.7 Conclusion

In summary, achieving the state of robonomics will require the hospitality industry to redesign all facilities to make them robot friendly. The challenge of achieving such a design relies on balancing the needs and abilities of robots with the aesthetic and functional preferences of humans, who will remain the customers of hospitality businesses. As the hospitality industry transitions to the state of robonomics, it is likely to see the emergence of new business types (e.g. plug-and-play hotels) and changes in operating procedures that will create new demands for the facilities' design. Furthermore, all of the emerging changes will need to be supported by the overarching system infrastructure to enable connectivity and communication across all robots, smart devices and smart surfaces to ensure a supportive and friendly environment for robots.

The design should follow the latest developments in technology, be fluid and easily transformable to support robots' duties and create inclusive environments for guests too. Robots in the facility should be kept 'informed' about changes in the design in order to perform their tasks efficiently. Smart technologies utilised in the facility design will not only support robotic employees but also contribute to sustainability efforts, economic effectiveness and improved customer experience.

Big corporations may lead the way by developing all of these new changes in design by establishing partnerships with manufacturers and start-up companies. Therefore, we forecast a tight partnership between big hospitality companies, technology leaders, suppliers, manufacturers and governments. It is commonly known that big corporations sometimes find it hard to innovate. However, due to high initial entry costs, they are the ones who have enough capital to innovate, grow and lead the industry. Consequently, mergers and acquisitions will probably be seen between hospitality, technology and manufacturing companies too. These first movers will create oligopolies, where they will have the greatest share of the market while independent hotels will follow and, therefore, will get a smaller share.

One of the most important elements of this partnership should be interoperability, such as the Hospitality Technology Next Generation (HTNG), which is an organisation that develops standards for the industry and helps companies to adopt optimal designs for their systems. This interoperability would enhance the connection between robot suppliers/manufacturers and hospitality companies, setting standards that will facilitate coordination and operability. At the same time, all these new changes should be regulated by governments at the national and other levels. They should be aware of these changes and should continue to update the regulations to meet the realities of the state of robonomics. When more and more tasks are automated and performed by robots, our society will require new regulations for the navigation of autonomous vehicles, accident prevention and insurance; digital privacy and the protection of consumers; fair use of artificial intelligence; occupancy guidelines accounting for both humans and robots; and other norm-regulating and safeguarding human–robot interactions. Such changes will facilitate the incorporation of robots as a part of the labour force and guests in the hospitality industry and regulate their use and interaction with humans.

Building regulations should also provide directions for safety and accessibility guidelines. Regulations on the floor and elevator robot capacity; navigation priority rules; evacuation rules: people first, then service robots (assisting people with disabilities/needs) then robotic guests, following essential robotic employees, and finally other robotic employees, should be discussed and regulated too; as well as emergency power generators and batteries capacity and use, just to mention some examples. Currently, hotels have regulations for how many parking spots they should have per room, but in a robonomics economy, authorities need to develop regulations for charging stations, drone landing space and so on.

At the same time, robotic vendors should work with hospitality technology providers and industry professionals to make sure that a robot may be integrated into the hospitality technology landscape. It is important for robotic vendors to understand which hospitality technology systems each type of robot needs access to. Industry professionals need to be aware of the infrastructure (physical and digital) demands placed on properties by employing robots and creating robot-friendly environments. Hospitality technology vendors need to be aware of the functions that are being delegated to robots vs the functions that are being kept under humans' oversight and develop such functions and underlying algorithms accordingly.

As the complexity of hospitality technology continues to grow, there will be a continuous need for consolidation in hospitality technology systems and the data that flows between them (Mogelonsky & Mogelonsky, 2021) and is accessible to robots, e.g. property management systems (PMS), point of sale systems (POS), electronic locking systems (ELS), revenue management systems (RMS) and SBFMS. Another potential approach to solving the challenge of system interoperability and data sharing could be to use open infrastructure that would allow hospitality providers to select technology systems from vendors they prefer to use and build 'stackable' technology solutions, communicating with each other through application programming interfaces (APIs).

All changes that have been discussed in this chapter would create an impact on the customer experience. The operational, systemic and architectural changes will affect the way in which hospitality facilities look and feel, how services are delivered and what technology infrastructure is put in place to support all these changes. While this chapter provided numerous examples of how hospitality facilities should be changed to become robot friendly, it is important to keep in mind that the primary customers of hotels are expected to be humans, even in the state of robonomics. Therefore, changes that are implemented with robotic employees in mind should be supportive of building positive experiences for human guests, leading us to the idea that robot-friendly facility design needs to facilitate human–robot interaction and be inclusive of both humans and robots.

References

Alexandris, K., Kouthouris, C. and Meligdis, A. (2006) Increasing customers' loyalty in a skiing resort. *International Journal of Contemporary Hospitality Management* 18 (5), 414–425.

Ali, F.Y. and Ali, S.A.D.H. (2020) The impact of nanomaterials on the dynamics of contemporary architecture. *IOP Conference Series: Materials Science and Engineering* 745 (1), 012121.

Ali, F., Omar, R. and Amin, M. (2013) An examination of the relationships between physical environment, perceived value, image and behavioral Intentions: A SEM approach towards Malaysian resort hotels. *Journal of Hotel and Tourism Management* 27 (2), 9–26.

Ali, F., Amin, M. and Ryu, K. (2016a) The role of physical environment, price perceptions, and consumption emotions in developing customer satisfaction in Chinese resort hotels. *Journal of Quality Assurance in Hospitality & Tourism* 17 (1), 45–70.

Ali, F., Kim, W.G. and Ryu, K. (2016b) The effect of physical environment on passenger delight and satisfaction: Moderating effect of national identity. *Tourism Management* 57, 213–224.

Atwa, M., Al-Kattan, A. and Elwan, A. (2015) Towards nano architecture: Nanomaterial in architecture – A review of functions and applications. *International Journal of Recent Scientific Research* 6 (4), 3551–3564.

Bare Conductive (2020) Smart surfaces will transform the way we live, work and care. See https://www.hackster.io/news/smart-surfaces-will-transform-the-way-we-live-work-and-care-200b6b8e7010 (accessed 27 May 2021).

Berezina, K. (2018) Hotels, get ready for robots. *HITEC Bytes 2018 Special Report*.

Berezina, K., Ciftci, O. and Cobanoglu, C. (2019) Robots, artificial intelligence, and service automation in restaurants. In S. Ivanov and C. Webster (eds) *Robots, Artificial Intelligence, and Service Automation in Travel, Tourism and Hospitality* (pp. 187–219). Emerald Publishing.

Bourée, W. and D'Andrès, L. (2018, December 3) How AI revolutionizes floor surface? Technis. See https://technis.com/news-coverage/how-ai-revolutionizes-floor-surfaces/ (accessed 3 June 2021).

Canton, J. (n.d.) Hotels.com™ Hotels of the future study. https://www.hotels.com/page/hotelsofthefuture/ (accessed 20 September 2020).

Cao, l. (2020) What to expect from interiors of the future. *Arch Daily*. https://www.archdaily.com/935089/what-to-expect-from-interiors-of-the-future (accessed 28 September 2020).

Cohen, S.A. and Hopkins, D. (2019) Autonomous vehicles and the future of urban tourism. *Annals of Tourism Research* 74, 33–42. https://doi.org/10.1016/j.annals.2018.10.009

Computing Community Consortium (2020) *A Roadmap for US Robotics: From Internet to Robotics, 2020 edition*. Computing Community Consortium. See https://www.cccblog.org/2020/09/09/robotics-roadmap-for-us-robotics-from-internet-to-robotics-2020-edition/ (accessed 22 September 2020).

Countryman, C.C. and Jang, S. (2006) The effects of atmospheric elements on customer impression: The case of hotel lobbies. *International Journal of Contemporary Hospitality Management* 18 (7), 534–545.

DroneTrader (2020, May 17) List of manned passenger drones and drone taxis. https://blog.dronetrader.com/top-passenger-drones-helicopters-drone-taxis/ (accessed 17 September 2020).

EHang AAV (n.d.) The era of urban air mobility is coming. https://www.ehang.com/ehangaav (accessed 17 September 2020).

Elara, M.R., Rojas, N. and Chua, A. (2014, May) Design principles for robot inclusive spaces: A case study with roomba. In *2014 IEEE International Conference on Robotics and Automation (ICRA)* (pp. 5593–5599). IEEE.

Han, H. and Hyun, S.S. (2017) Impact of hotel-restaurant image and quality of physical-environment, service, and food on satisfaction and intention. *International Journal of Hospitality Management* 63, 82–92.

Hanaysha, J. (2016) Testing the effects of food quality, price fairness, and physical environment on customer satisfaction in fast food restaurant industry. *Journal of Asian Business Strategy* 6 (2), 31–40.

Hotel Innovation Committee (2019, November) Smart technology guide 2019. https://www.stb.gov.sg/content/dam/stb/documents/industries/hotel/Smart%20Hotel%20Technology%20Guide%202019.pdf (accessed 23 September 2020).

Ivanov, S.H. and Webster, C. (2017a, June) The robot as a consumer: A research agenda. In *'Marketing: Experience and Perspectives' Conference* (pp. 29–30).

Ivanov, S. and Webster, C. (2017b, October) Designing robot-friendly hospitality facilities. *Proceedings of the Scientific Conference 'Tourism. Innovations. Strategies'* (pp. 74–81). Bourgas, Bulgaria.

Ivanov, S.H., Webster, C. and Berezina, K. (2017) Adoption of robots and service automation by tourism and hospitality companies. *Revista Turismo & Desenvolvimento* 27 (28), 1501–1517.

Ivanov S., Webster C. and Berezina K. (2020) Robotics in tourism and hospitality. In Z. Xiang, M. Fuchs, U. Gretzel and W. Höpken (eds) *Handbook of e-Tourism* (pp. 1873–1900). Springer. https://doi.org/10.1007/978-3-030-05324-6_112-1

Kim, W.G. and Moon, Y.J. (2009) Customers' cognitive, emotional, and actionable response to the servicescape: A test of the moderating effect of the restaurant type. *International Journal of Hospitality Management* 28 (1), 144–156.

Lockwood, A. and Pyun, K. (2019) How do customers respond to the hotel servicescape? *International Journal of Hospitality Management* 82, 231–241.

Löffler, D., Schmidt, N. and Tscharn, R. (2018, February) Multimodal expression of artificial emotion in social robots using colour, motion and sound. In *Proceedings of the 2018 ACM/IEEE International Conference on Human–Robot Interaction* (pp. 334–343).

Mazzara, M., Afanasyev, I., Sarangi, S.R., Distefano, S., Kumar, V. and Ahmad, M. (2019, June) A reference architecture for smart and software-defined buildings. In *2019 IEEE International Conference on Smart Computing* (SMARTCOMP) (pp. 167–172). IEEE.

Mogelonsky, L. and Mogelonsky, A. (2021) Consolidation of data is the next big step for hotels. *Hospitalitynet*. https://www.hospitalitynet.org/opinion/4107118.html (accessed 1 November 2021).

Nanu, L., Ali, F., Berezina, K. and Cobanoglu, C. (2020) The effect of hotel lobby design on booking intentions: An intergenerational examination. *International Journal of Hospitality Management* 89, 102530.

Palmer, A. (2020, August 31) Amazon wins FAA approval for Prime Air drone delivery fleet. https://www.cnbc.com/2020/08/31/amazon-prime-now-drone-delivery-fleet-gets-faa-approval.html (accessed 17 September 2020).

Pinillos, R., Marcos, S., Feliz, R., Zalama, E. and Gómez-García-Bermejo, J. (2016) Long-term assessment of a service robot in a hotel environment. *Robotics and Autonomous Systems* 79, 40–57.

Ryu, K., Lee, H.R. and Kim, W.G. (2012) The influence of the quality of the physical environment, food, and service on restaurant image, customer perceived value, customer satisfaction, and behavioral intentions. *International Journal of Contemporary Hospitality Management* 24 (2), 200–223.

Simpeh, K.N., Simpeh, M., Nasiru, I. and Tawiah, K. (2011) Servicescape and customer patronage of three star hotels in Ghana's metropolitan city of Accra. *European Journal of Business and Management* 3 (4), 119–131.

Smart Hotel Technology Guide 2019 (2019, November) The Singapore Hotel Association and the Singapore Tourism Board. https://sha.org.sg/publications/smart-hotel-technology-guides (accessed 24 May 2023).

Song, S. and Yamada, S. (2017, March) Expressing emotions through color, sound, and vibration with an appearance-constrained social robot. In *2017 12th ACM/IEEE International Conference on Human-Robot Interaction (HRI)* (pp. 2–11). IEEE.

Stone, P., Brooks, R., Brynjolfsson, E., Calo, R., Etzioni, O., Hager, G. ... Leyton-Brown, K. (2016) *Artificial Intelligence and Life in 2030. One hundred year study on artificial intelligence: Report of the 2015–2016 study panel*. Stanford University, Stanford, CA. http://ai100.stanford.edu/2016-report (accessed 22 September 2020).

Tan, N., Mohan, R.E. and Watanabe, A. (2016) Toward a framework for robot-inclusive environments. *Automation in Construction* 69, 68–78. https://doi.org/10.1016/j.autcon.2016.06.001

Treptow, A., Masselli, A. and Zell, A. (2004) Real-time object tracking for soccer-robots without color information. *Robotics and Autonomous Systems* 48 (1), 41–48. https://doi.org/10.1016/j.robot.2004.05.005

Tutuncu, O. (2017) Investigating the accessibility factors affecting hotel satisfaction of people with physical disabilities. *International Journal of Hospitality Management* 65, 29–36.

United States Access Board (n.d. a) About the rulemaking on public rights-of-way. https://www.access-board.gov/guidelines-and-standards/streets-sidewalks/public-rights-of-way (accessed 17 September 2020).

United States Access Board (n.d. b) Guidelines and standards. https://www.access-board.gov/guidelines-and-standards (accessed 17 September 2020).

Ward, T. (2020, September 9) Walmart now piloting on-demand drone delivery with Flytrex. https://corporate.walmart.com/newsroom/2020/09/09/walmart-now-piloting-on-demand-drone-delivery-with-flytrex (accessed 17 September).

Webster, C. and Ivanov, S. (2020) Future tourism in a robot-based economy: A perspective article. *Tourism Review* 75 (1), 329–332. https://doi.org/10.1108/TR-05-2019-0172

9 Sex, Health and Wellness: Considering the Future Potential for Robots and Human Relationships within Tourism Resorts

Daniel Wright

9.1 Introduction

> Humans in the present day seem obsessed with robots, real and imagined, as we embrace duelling visions of robo-utopias and robo-dystopias that titillate, bring hope, and scare the bejesus out of us (Duncan, 2019: 4)

As society moves further into the 21st century, we are on the cusp of the fourth Industrial Revolution. The first Industrial Revolution (end of 18th and beginning of 19th century) saw the industrialisation of production. Here, the manufacturing and production of textiles moved from people's homes to factories. During this period, steam power played an important role. The second Industrial Revolution (end of 19th and beginning of 20th century) was dominated by developments in energy, such as electricity, gas and oil. This led to revolutions in transport via combustion engines and developments in aviation, infrastructure with steel demand, medicine due to chemical synthesis and communication via telegraph and telephones. The third Industrial Revolution, the digital one, emerged around the middle of the 20th century and has been driving humanity through a range of technological changes, which have transformed the way we live and who we are as a species. Here, we saw the rise of nuclear energy, electronics, telecommunications and computers. The fourth Industrial Revolution (also known as Industry 4.0) is characterised by a fusion of technologies that will blur the lines between the physical, digital and biological spheres (Schwab, 2016). Part of this should see the rise of robotic technology. As noted in the above quote by Duncan (2019), human fascination with robots is dynamic and likely to continue during Industry 4.0. How and why are important questions, and this chapter considers potential future scenarios from a health, wellness and leisure perspective.

When examining and presenting future worlds, visions can project non-fictional content, from academia, within industry and by governments (Caraian *et al.*, 2015; Ivanov & Webster, 2018; Ivanov *et al.*, 2019; Yeoman & Mars, 2012). Likewise, fictional forms within popular culture offer depictions of technologically machine-driven future worlds showcasing more often than not dystopic ideas (*Metropolis*; *Blade Runner*; *The Terminator*; *WALL-E*). Utopian images of robots in the future are less commonly explored in fiction. As noted by Larson (2002), when considering alternative futures, it is apparent that negative and positive scenarios are often not symmetric to each other, and Larson suggests that negative scenarios are often easier to formulate. According to Larson (2002), it is more simplistic to extrapolate current trends and present scenarios where our world is moving towards disorder and crisis. Contrary to shocking (fictional) worldly illustrations where machines have taken control, the fourth Industrial Revolution does not need to lead humanity down a dystopic path. Research (such as that presented above) is proving that incorporating artificial intelligence (AI), technology and robotics could prove useful to human beings (e.g. see Carnevale, 2015). Attempts have been made to showcase futures where machines become integrated with humans in their social spaces, working together to ensure prosperity, innovation and successful collaborative achievements for the longevity of not only humans, but also the machines and our natural environment.

A key challenge is establishing and presenting future scenarios. What will society be like in 2068? For people who grew up during the 20th century, early 21st-century society was depicted to be a place of flying cars, robot worlds and humans being beamed into space. However, as we now know, this is not our current reality. Visioning the future by exploring current trends and how and when they could evolve and develop over time is complex. Thus, offering predictions and scenarios of the future is a somewhat complicated task, but necessary. Individuals, businesses, organisations and governments take part in the act of planning for the future. In doing so, society aims to better prepare for (uncertain) future times and, significantly, aims to create a healthier environment and improved social conditions.

A key part of who we are as a global community revolves around our desire to grow and advance who we are individually and collectively, and this includes our work and leisure time. These actions often coincide around travel and our interactions with other people. Work and leisure practices have changed and evolved throughout human history and, more recently, our industrial revolutions have played a significant role. As we move further into the 21st century and Industry 4.0 potentially contributes to our progress, it is necessary to consider how we could evolve as a species. How will we aim to improve and grow individually and collectively? Seeing that travel forms a significant role in our leisure experiences, what forms could become available because of the new Industrial Revolution? In line with this, the chapter explores how android robotic technology could actively operate within the tourism and leisure industry in the future. Specifically, two scenarios are presented, exploring how humans and androids could interact in future tourism settings. Scenario 1 explores the idea of android robots becoming part of the sex tourism industry. Scenario 2 focuses on health and well-being within the travel industry and considers future tourists interacting with android therapists. However, it must be stressed that this chapter is not advocating the future scenarios presented here, instead it aims to shine a light on past and present trends in society as a means to highlight where we could be heading in the future, in this case, around 2068. In doing so, the chapter challenges the reader's perspective of current technologies and where they could lead us in the future.

9.2 Humanoid, Android and Gynoid Robots

Many aspects of our lives once carried out by humans are now performed by machines (Allain, 2015). As noted by Bahishti (2017), robotisation is a process in which tasks (once carried out by humans) are now performed by machines. Such machines are mechanical or software-based applications, often referred to as automation. However, 'among those advancements, robots have become significant by managing most of our day-to-day tasks and trying to get close to human lives. As robotics and autonomous systems flourish, human-robot relationships are becoming increasingly important' (Bahishti, 2017: 60). Robotisation is not something that will happen in the future, it started many years ago. Cresswell *et al.* (2018: 1) suggest that 'the emergence of robotics is transforming industries around the world. Robot technologies are evolving exponentially, particularly as they converge with other functionalities such as AI to learn from their environment, from each other, and from humans'.

The initial idea of robots is arguably a non-human form and genderless. However, as humans become more accustomed to witnessing more human-like robots, it is evident that more (non)gender-specific robots will take shape. Science fiction has presented society with a range of different types of artificial creations that resemble humans. Two terms, humanoid and android, are often used within the literature. However, there is a difference between them (see Table 9.1). The difference, which is explored by Brown (2020), points towards the level in which they (the two types) mimic humans and subsequently the emotional reaction that people have towards them. A humanoid robot's body takes the form of a human body with a head, arms, legs, a torso (on some occasions the legs

Table 9.1 Difference between humanoids and androids

Characteristic	Humanoid robots	Android and gynoids
Definition	Robots made in the form/shape of a human body – including a head, torso, two arms and two legs	Artificial beings that resemble a human in external appearance and behaviour
Composition	Often made from metal or other non-organic material	Composed of any material including organic matter
Human aspect	Human motion/basic or simple behaviour	Human aesthetic appearance
Purpose or goal	Performs human tasks and simulates human movement	Looks human-like/indistinguishable from humans as possible
Application	Replace humans in various tasks and allow researchers to study human movement	Cognitive, social and neuroscience to better understand human behaviour and interaction
People's perception and emotional reaction	No emotional reaction, or as much emotion as for a machine	Evokes fascination as well as discomfort and eeriness
Human replacement	Replacing human utility – such as daily tasks and activities	Replacing human relationships and interaction

Source: Adapted from Brown (2020).

might be replaced by another form, such as wheels). Humanoid robots are made of non-organic material, often metal. Humanoid robots are created to mimic human motion and behaviour and as their capabilities improve, so does their usefulness. Their role in society is becoming more apparent because they can take on difficult, dangerous and repetitive tasks better than humans. Humanoid robots are the ones that society fears most not only because of discussions on robots taking over human jobs and making humans obsolete (Brown, 2020), but also because of their appearance. This idea was explored by Mori (1970), termed 'the uncanny valley'. Mori (a roboticist) noted that there is a point in which robots begin to look more like humans and project a sense of eeriness. Androids, like humanoids, are an artificial being that aim to look aesthetically human like. The term android was originally used as a genderless humanoid robot, but in the Greek prefix 'andr' refers to man, in the masculine-gendered sense (Liddell *et al.*, 1996). In her novel the *Divine Endurance*, Gwyneth Jones (1985) used the term gynoid, here describing a robot slave, a character in futuristic China who is judged by her beauty (Tatsumi, 2006). Currently, androids are defined on the level of the technology from which they are constructed, which is mainly a mix of internal robotic mechanisms and a lifelike exterior, often a synthetic (both organic and non-organic) flesh-like material. The main aim of androids is to mimic the human appearance. Unlike humanoids, androids are not necessarily robotic or mechanical. Using synthetic flesh allows androids to mimic facial expressions and therefore a human likeness. Thus, unlike humanoids, androids and gynoids are designed to be more 'indistinguishable' from humans, especially from their external appearance and behaviour (Brown, 2020).

While android robots still lack the ability of those portrayed in science fiction hits such as *Star Wars* and *Blade Runner*, and none have yet mastered walking on two feet, the current level of technology showcases how far society has come. Current innovative examples provide us with an insight into the present state of the technology available to us. As we move further into the future, android robots are becoming more human like. There are many innovative examples, including creations by Hiroshi Ishiguro which comprise of the Geminoid HI-1 (see Stafford, 2016) and Geminoid F (see Rogers, 2021). Another android named Sophia (created by Hanson Robotics) can carry out a wide range of human actions and is capable of making up to 50 facial expressions and can equally express feelings (Dang, 2019; Tangermann, 2017). Sophia has very expressive eyes, her AI revolves around human values and she can express a sense of humour. Sophia was introduced to the United Nations on 11 October 2017 and on 25 October she became the world's first robot citizen when granted Saudi Arabian citizenship, making her the first android robot to have a nationality (Dang, 2019). Another example is Erica, developed in Japan with a special emphasis on her speech capabilities. While Erica cannot walk, she is able to interact easily with human beings and change her facial expressions according to the conversation taking place. Currently, she cannot move her arms, but Erica can move her facial features, neck, shoulders and waist independently. According to Hiroshi Ishiguro, Erica has a 'soul' but that is a statement that creates a debate more on robotic metaphysics and less on technology (Specktor, 2018). These examples provide an insight into the current level of technology around robots and, importantly, it offers a visual springboard into what the future could hold.

As we move further into the 21st century, current research into robotics and AI will continue to develop new machines, ones that not only perform repetitive tasks, but can also learn and adjust their behaviour to interact with people, working together as partners (Bahishti, 2017). According to Merrifield (2013), these robots are designed to learn and adapt to changing circumstances, interacting with society in socially appropriate ways by interpreting human needs and intentions. Machine learning could provide the necessary platform for android robots to start learning from their own mistakes and eventually correct them, adapt and improve. The future looks promising for android robots and their eventual ability to become more like humans. Coincide this with machine learning and androids could become even more human like, as they start to learn from their own environment and adapt to it. However, ethical questions do arise from this. As noted by Chatila (2018,) when robot designers seek perfect human mimicry and blur the lines between humans and machines, problems arise. According to the authors, the following three issues need to be considered: (1) human identity and human dignity; (2) human perception of androids; and (3) actual or potential applications using androids, or for what do we use physical androids. These are considered when exploring the potential for androids in tourism in the following two scenarios.

9.3 Tourism Scenarios for 2068

Having considered android robotic technology and its increasing fusion with AI and machine learning capabilities, the following section now presents two scenarios in which android robots could interact with future tourists. Predicting the future can be defined as forecasting the future on a systematic basis by studying current-day trends in human affairs (The Oxford English Dictionary, 2015). Thus, acknowledging and justifying the future should include an analysis of past and current 'trends' (Wright, 2016). Predicting the future is no easy task, and many predictions turn out to be incorrect. However, this should not discredit the value of doing so. As Friedman (2009) notes that he does not have a crystal ball, but a tried and tested method in anticipating the future involves the exploration of current events, trends and technological developments. The approach taken here in exploring the future is via the presentation of scenarios. Scenario planning is a commonly applied method when exploring the future; by presenting stories with narratives (Lindgren & Bandhold, 2009), offering a setting, a time frame and additional details commonly associated with storytelling, allowing an audience to further engage and visually explore the ideas being presented (Wright, 2019). Schwartz (1991: 4) notes that scenarios can be used as a 'tool for ordering one's perceptions about alternative future environments in which one's decisions might be played out. Alternatively: a set of organised ways for us to dream effectively about our own future'. Each of the two scenarios are followed by a scenario discussion section, offering further justification for and consideration of the potential eventuality of the scenario in the future and how this could impact on the tourism and hospitality industry.

9.3.1 Scenario 1 – Android pleasure resort: A place for tourists to explore their sexual fantasies

In 2068, tourists will travel to different resorts around the world to understand and satisfy their gender needs and to indulge their sexual fantasies and curiosities, and they will do so with android and gynoid robots. By 2068, android and gynoid robots will aesthetically be human-like. The robots will look and act like humans. The robots will be able to operate independently due to advancements in machine learning and will engage with humans on a sociable and intimate level. International tourism resorts will provide tourists with a range of personalised experiences, from simple communicative interactions to sexual intimacy. The more money tourists are willing to spend, the more extravagant their personalised experience can become. Gender identity and sexual orientation evolved during the 21st century to become more diverse than ever before. In the resorts, humans can simply spend time having intimate discussions with androids and gynoids to having intimate sexual encounters (with more than one). Tourists can even (pre)request a personal robot design (tailor-made to their own preferences), including past and present individuals and celebrities (personalisation comes at a higher cost). The resort has something to offer all gender types and people interested in a range of sexual orientations. Androids and gynoids in the resorts operate as support robots to humans.

9.3.2 Scenario 1 discussion

Technology has already influenced our engagement with and understanding of physical and emotional intimacy and sexual encounters individually and with others. Be it dating apps, porn sites or sexual toys. People use technology as a tool to locate sexual encounters or to have/enhance sexual experiences. More recently is the emergence of techno-sexuals living with life-sized dolls. Society is potentially at the dawn of a new sexual revolution, and with many people already angered by the changes, it is time to face our reality, where technology may take the most intimate area of our lives (The Medical Futurist, 2018). 'Sex robots threatening the world's oldest profession. Technosexuals living with life-sized dolls. At the dawn of a new sexual revolution, it's time to face where technology may take the most intimate area of our lives' (The Medical Futurist, 2018). The discussion around sex with robots is one that has been explored in many disciplines. Pearson (2015) suggests that sex rarely requires an introduction, but what is required are conversations around the ways one might grow to enjoy it and how it could evolve significantly in coming years. Unsurprisingly, the matter of sex robots is a controversial one. A sex robot can be described as a mechanical doll with a human-like appearance and size that is able to perform intercourse by means of motor-equipped artificial genitalia (Choi, 2008; Pearson, 2015; Richardson, 2016). According to Appel *et al.* (2019), while a high degree of autonomy or AI may not be mandatory, more advanced prototypes of sex robots are capable of carrying out simple verbal communication, simulate different personalities and express enjoyment in order to increase their anthropomorphic realism. Clearly there are ethical issues, with scholars expressing concerns around objectification, gender stereotypes and non-empathetic forms of sexual encounter (Richardson, 2016; Sullins, 2012). Contrastingly, others see potential psychological and social benefits of sexual robot partners. These include sexual therapy opportunities and sex robots as an alternative to human sex workers (Devlin, 2015; Levy, 2008). While it is uncertain when sex robots will be available, some have forecasted that high-income households will see them by 2025 and that human–robot sex will overtake interpersonal sex by 2050 (Levy, 2008; Pearson, 2015).

Travel and sex were commonplace with our ancestors, as it is today, and will likely be in the future. 'Men of antiquity worshipped Aphrodite by having sex with her temple maids at Paphos on Cyprus' (Watson, 2016). Watson (2016) notes that, in the pre-Christian era 'hoards of pilgrims who came to her shrine would evoke her name before having intercourse with her temple maids. It was an ancient form of sex tourism'. According to IAMAT (2017), sexual tourism continues to be a very lucrative industry spanning the globe. ECPAT International estimates that each year, approximately 250,000 people travel internationally to engage in sex tourism and that the industry generates over US$20 billion in revenue (IAMAT, 2017). Goldapple (2020) notes that today's sex toy market is worth a predicted US$45 billion, and is advancing rapidly as companies apply robotics to sex dolls. The idea of human sexual encounters with robots is not a new one, nor is it within the context of tourists engaging with sexual robot encounters in the future. Yeoman and Mars (2012) presented a scenario which discussed the potential for a robot sex industry in Amsterdam's red-light district, where clean android prostitutes could satisfy men in the year 2050. Currently, tourists to Barcelona can savour the world's first hyper-realistic sex doll brothel (Lumidolls, 2021). However, as we move further into the 21st century, how will sex tourism evolve? Could sex move in to resorts? According to Hutton (2019), from an Asian perspective, sex tourism is changing, and travellers are seeking more adult-only resorts. Tourism resorts are commonplace, and the likelihood of androids and gynoids wandering around cities indecently in 2068 is questionable. Therefore, the scenario focuses on the idea that such practices will also exist in controlled and managed tourism resorts.

Many of the challenges will depend on how society in the future interprets sexual orientation. The term gender binary describes a system in which a society allocates members into one of two sets of gender roles and gender identities and allocates people based on their type of genitalia (Lorber & Moore, 2010). According to the World Health Organisation (WHO), gender refers to the characteristics of men, women, boys and girls that are socially constructed. Furthermore, these include norms, behaviours and roles associated with being a man, woman, boy or girl, as well as relationships with each other. As gender is a social construct, gender varies from society to society and can change over time. While gender and sex are related, they are different to gender identity. According to the WHO (2021), gender identity 'refers to a person's deeply felt, internal and individual experience of gender, which may or may not correspond to the person's physiology or designated sex at birth'. Traditionally, people would say they have a sexual orientation; however, as noted by Selterman (2015), human sexuality is not simple and straightforward, and in today's society, with ongoing changes in what it means to have a specific gender, discussions are becoming more complex. Sexual preferences are not set in stone and can change over time, often depending on the immediate situation of the individual. This has been described as sexual fluidity. Another related concept is that of erotic plasticity, defined as changes in people's sexual expression (attitudes, preferences and behaviour). Importantly, someone's sexual response can fluctuate depending on their surrounding environment (Selterman, 2015). Ultimately, gender, sexuality and sexual preferences are a tricky and complex arena. What is important for this chapter is the recognition of this and how this could evolve and change in the future and how it could impact the leisure market (such as hotels and resorts). If humans are socially constructed and our attitudes, preferences and behaviour can change, as can our genders, then our desires and curiosities are likely to be different in the future. In the future, a tourism robot sex resort, such as the one presented in the above scenario, could offer experiences for a spectrum of human gender and sexual orientation. This alone raises many interesting and challenging questions around the management (supply) of such resorts and likewise the touristic demand for such resorts.

From a supply perspective, there will be issues around the investment and upkeep of sex robots if owned by the resort. Resorts will have to operate under governmental (regional, national and international) regulations and laws, many of which today still do not exist when discussing robots. However, with greater development in future robots (see Erica and Sophia above) who could have self-dependency, individual identity, citizenship of a country (and even a soul), then questions need to be explored around the potential treatment of them. As noted, future robots could be owned by the resorts or travel independently (or with a human or another robot partner). How are they treated, like humans (with the same rights)? How much do you charge a robot staying at your resort? Do they need food? What about charging and power? Of course, these are not easy questions to answer, because at present they are just that, questions that need consideration, especially if the above scenario was to become a reality. The following model aims to provide a visual consideration to the internal and external environment of a future resort as presented in the above scenario (Figure 9.1). The model considers the management of the internal environment, which would include the physical upkeep, maintenance and development of the resort. Likewise, the management of employees and the requirements of owners and shareholders (which in the future could be either human and/or robot). From an external perspective, consideration would have

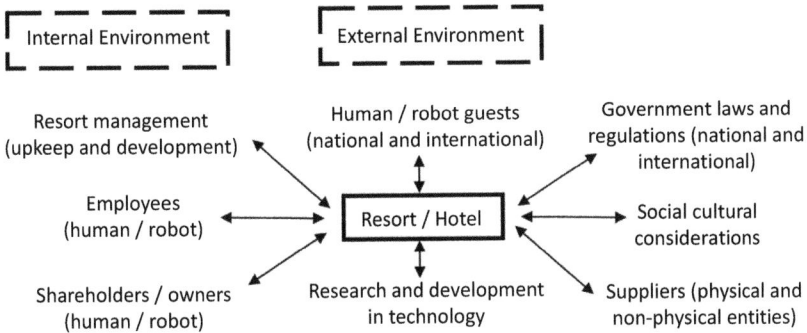

Figure 9.1 Model 1: Internal and external business environment for future sex resort/hotel

to be given to wider changes in laws and regulations passed by government, social-cultural changes and research and development in future technologies. Additionally, the external environment would consist of potential changes in the suppliers of physical business entities (such as produce, products and potential sex robots) and non-physical entities such as learning new skills – i.e. sex robot legislation.

9.3.3 Scenario 2 – A future health and wellness centre: Run by robots and human therapists for robot and human guests

In the future, guest/tourists (which could consist of robots, humans or transhumans) could travel to health and wellness centres and spend time engaging with robot (android, gynoid and transgendernoid) and human therapists. As humans (and potentially robots) seek health and wellness guidance, future centres could see a greater balance between future robots and humans taking an active role in providing guest with the ultimate resolutions in seeking the best solutions to personal physical and mental well-being. As society moves further into the 21st century, there will be a greater balance between robots and humans as they take up their place/roles in society. Future health and wellness centres with human and robot therapists will offer a variety of different cultural and environmental settings. Humans will continue to endure mental and physical challenges throughout their lifetime, and therefore, will continue to seek alternative centres (and different types of therapists) for support and relief. Robots could become a new type of tourist, as they become integrated into society, but overtime, like humans seek mental and physical support. Health and wellness travel will continue to grow in importance, popularity and variety throughout the century. Tourists will consist of not only humans but also robots, and they will seek not only traditional healing methods, but also support, advice and guidance from different types of therapists – robot and humans.

The role of the robot therapist: Humans (of all ages) could feel comfortable around robots as they become human-like and can read and understand humans (their body language, facial expressions and psychological state of mind) and adapt to their physical presence and emotional state. Robots can draw on greater knowledge than human therapists, as robots are hard-wired to the internet. Likewise, robots have access to patient's digital data. Humans have been imputing all their data into the digital cloud, through mobiles devices, smart watches, augmented reality glasses (lenses) and other digitally connected devices, including but not limited to leisure and sports items (e.g. bikes and tennis rackets) and household items (e.g. cooking appliances and fringes). All this data provides the robot therapist with a vast amount of data on the individual (both mental and physical and in relation to one's leisure and work activities and how the individual feels towards them). With instant access (via the internet) to all human knowledge surrounding mental and physical wellness (from different schools of thought, from sociology, psychology, science and medicine, spirituality and religion) and with the personal data of the individual, the robot is best equipped to provide the most suitable diagnosis and solutions to the human. However, they might have their limitations in carrying out physical wellness activities – here, humans could still have a role to play.

The role of the human therapist: Humans will still play an active role in the future and offer support not only to humans but also to robots in wellness centres. As the creators of robots, humans are important and even godlike to robots, as their wisdom and evolutionary prowess would be regarded and valued by robots in their evolutionary journey. Like humans, robots will succumb to the challenges of life and the difficulties of growth and placement in society. Therefore, health and wellness centres will act as a place for them to seek mental and physical support. Like humans, robots will have the choice as to who best can meet their different needs. Much of this will come down to the capability and expertise of the (robot and/or human) therapists at the resorts and likewise the resorts' approach to managing wellness development.

9.3.4 Scenario 2 discussion

There is no definitive answer or definition for health, wellness and medical tourism within the literature. What currently exists is a range of terms, categories and meanings from a range of authors. This is highlighted by Fetscherin and Stephano (2016), who recognise the somewhat loosely and poorly categorised understanding of various terms such as health tourism, medical tourism and wellness tourism. Smith and Puczkó (2014) suggest that medical tourists are driven to travel for treatment abroad, while 'health tourism' is composed of 'wellness tourism' and 'medical tourism'. Ghosh and Mandal (2018) agree that travel for physical well-being has long been a practice, while medical tourism is seen as an emerging phenomenon that is characterised by medical care combined with holiday-making activities. Connell (2006) notes that 'medical tourism' is a term applied when considering cases of medical, surgical or dental interventions, while all others are categorised as 'wellness tourism'. The focus in the above scenario is on health and wellness centres and not places for medical treatment. According to Smith and Kelly (2006), wellness tourists are concerned with seeking a physical and/or mental transformation in the self. Wellness tourism is often associated with health-related experiences, self-pampering and positive change (Smith & Puczkó, 2014). Cook (2010) identifies the importance of tourists travelling for wellness experiences that are not available in their home environment. Mueller and Kaufmann (2001) propose that wellness draws attention to the individual's own responsibility for their health and the importance of becoming aware of and making choices to support a healthier existence. Koskinen (2019) suggests that preventative and proactive aspirations associated with health are often linked to wellness tourism. According to the Global Wellness Institute (2018) report, the global wellness economy was worth a $4.5 trillion market in 2018, while the industry grew by 6.4% annually from 2015 to 2017, from a $3.7 trillion to a $4.2 trillion market. The above scenario presents the idea that tourists of the future will continue (as in the past and present) to seek improved health and wellness support in travel resorts. Unlike before, however, a new form of support will be available: robot therapists.

'People have been shown to be more compliant when a robot asks them to do something as compared with a person' (see Open Access Government, 2019). How will people interact with robots in the future and will they listen to them? There is limited research on this matter. However, findings by Saunderson and Nejat (2021) provide some insight. In their study, they found that people tended to respond better to robots when in a more social setting and a peer-to-peer role, rather than taking an authoritative role. Additionally, the robot was more persuasive when offering rewards over penalties. Research within the field of AI robots acting as therapists remains somewhat innovative. However, research has begun to explore the idea. Fiske et al. (2019) note that AI-supported virtually embodied psychotherapeutic devices are currently developing and emerging rapidly (through mobile and internet devices people can communicate to virtual doctors and therapists). The scenario above focuses on AI android (robot) therapists in the future (not virtual therapists). The development of AI robots is entering a new stage where the focus is on interaction with people in their daily environments. With the improvement in increasingly complex robots for use in rehabilitation, heathcare services and/or other applications, robot–human interaction is a rapidly growing area of research (Durães et al., 2018). Currently, the integration of AI in embodied agents is still at an early stage in mental healthcare, and is mainly used in psychotherapeutic practices and supporting emotional, cognitive and social processes (Eichenberg & Küsel, 2018). Recently, intelligent robot animals such as Paro (a fuzzy harp seal) and eBear have been used to support patients with dementia. They are classed as 'companion bots' and engage individuals as in-health care assistants, supporting elderly, isolated or depressed patients through companionship and interaction (Fiske *et al.*, 2019). AI robots are also proving successful with engaging children with different disorders, such as autism. Children have proven to react positively to robots as well, with Kaspar robot showing positive signs in educational support and social skills (Huijnen *et al.*, 2017). More in line with Scenario 1, Fiske *et al.* (2019) point out that AI-enabled robots are also entering the field of human sexuality as a form of sexual therapy and support, with companies offering adult sex robots (such as Roxxxy, who can speak, learn their human partner's preferences, register touch and provide a form of intimate companionship).

Like Scenario 1, ethical concerns are evident at this early stage. Fiske *et al.* (2019) suggested that there is still little information on how humans are affected by contact with therapeutic AI. Additionally, individuals in therapy are more vulnerable, and their interactions with robots could make them even more vulnerable due to their desire for care and support. Human emotions, thoughts and feelings could be transferred to a robot, and while a human therapist is able to understand how to process these due to the nuances involved, the algorithm of an AI therapist needs to be complex to sufficiently engage in therapy. Likewise, at this stage it is too soon to understand how such emotional input would impact on the AI's mind (see Open Access Government, 2019). Cresswell *et al.* (2018) note that robots that aim to reduce loneliness or offer emotional comfort, run the risk of creating an emotional dependency on the patients they support. Significantly, this is identified as a serious concern in relation to long-term use of AI interventions. AI (at least for now) is not supposed to act as a permanent presence in people's lives, but as a tool for encouraging and supporting change in behaviour. Significantly, as we move further into the 21st century, we are likely to see 'the integration of AI devices into our everyday lives and medical care is undoubtedly changing social expectations and practices of communication' (Fiske *et al.*, 2019: 6).

As pointed out in Scenario 2, a tourist could travel to different corners of the world to seek out different types of AI-embedded robot therapists, because 'ideas around embodied AI are culturally and historically shaped' (Fiske *et al.*, 2019: 7). Consequently, Scenario 2 presents the idea that different cultures and environments will (like the past and present) offer different culturally relevant support and ideas around physical and mental health. However, unlike the past and present, culturally relevant and diverse robot therapists could be part of the health and wellness

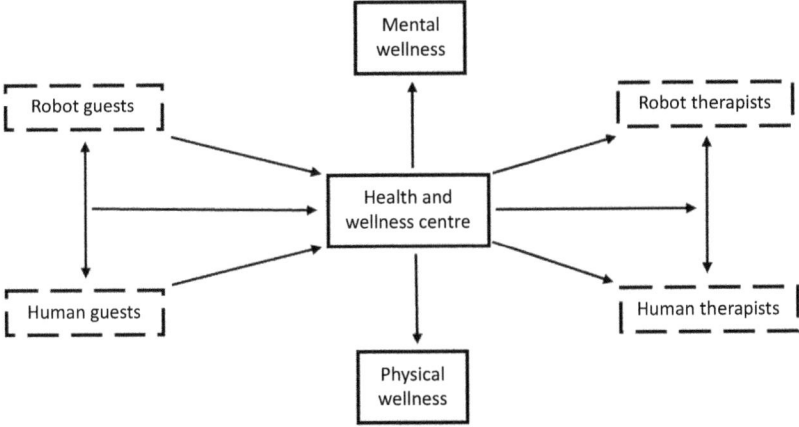

Figure 9.2 Model 2: Health and wellness centre: Human to robot guest and therapists

travel industry, alongside their human counterparts or as the dominant force. In future (health and wellness) travel resorts, tourists will experience and interact with the most data-driven and informed therapist humanity has even known. Here, the idea is that, once again, society has managed to integrate robots to assist and support humans. In order to envisage this, Model 2 provides a visual consideration of the relationship and choice between guest/tourist, a future health and wellness centre and the therapist (Figure 9.2). Going deeper, the challenging questions for the future in the model pose the following: Will the guest be a human or a robot or somewhere in between, i.e. transhuman? When visiting a health and wellness centre, are they seeking mental and/or physical wellness and guidance? Importantly, what type of therapist will be best placed to support them – robot, human or, again, something transhuman?

9.4 Concluding Remarks

As noted by Schwab (2016), 'the Fourth Industrial Revolution, finally, will change not only what we do but also who we are'. Moreover, it will impact our identity and our consumption patterns, it will shape our work and leisure time, and how we interact with others and the world around us. A key discussion around Industry 4.0 is the inclusion of robots. As robots become more intelligent, society will have to consider their role among humans, ideally aiming to map their abilities alongside human values (McDonald, 2015). As humans, we are creating the technology, thus we have the power to guide its evolution. Therefore, we should 'grasp the opportunity and power we have to shape the Fourth Industrial Revolution and direct it toward a future that reflects our common objectives and values' (Schwab, 2016). To achieve this, comprehensive, globally shared views on technology and society are likely to be needed and these could reshape our economic, social, cultural and human environments. As Schwab (2016) comments, 'in its most pessimistic, dehumanized form, the Fourth Industrial Revolution may indeed have the potential to "robotize" humanity and thus to deprive us of our heart and soul. But as a complement to the best parts of human nature – creativity, empathy, stewardship – it can also lift humanity into a new collective and moral consciousness based on a shared sense of destiny'. However, this postulates towards a global approach and potentially devalues the cultural and environmental distinctiveness of cultures and societies, something that is often seen as fundamental to travel and leisure. After all, we are often drawn to destinations as a result of their uniqueness and 'otherness' to what we can attain from our own home environments.

As noted, by offering future scenarios such as the two presented in this chapter, ideas of how robots and humans might interact challenge our current understanding. Ultimately, with increased human and robot interaction in the years to come, society will be able to look back and ask if enough was done to understand the development and impact, not just of robots (and their place and role in society), but also of wider societal changes and the impacts they are having on humans. Consequently, if these scenarios became reality in the future, how will future generations look back on societies' role today in shaping and developing the future. If robots and humans are integrated in a manner that humans are supported by robots, to better their lives and well-being, then we may be considered pioneers of a new world. If robots become an unwanted reality for future generations, then scenarios like this should act as a warning sign to current society, emphasising the need to give more time and attention to managing and regulating its rapidly progressive nature.

Bahishti (2017) notes that government departments (parliaments/governing bodies) are paying limited attention to special laws and regulations around robots. Nevertheless, as robots (and in this case, androids, gynoids and transgendernoids) become more human-like and intelligent, and play an increasingly active role in human life, it's only a matter of time before more legislative consideration is assumed. However, unlike the position and idea taken by Schwab (2016), who calls for a more global approach, maybe countries should take their own approaches (in terms of R&D and legislation). Maybe robots should be allowed to evolve in different ways, as humans have done, rather than approaching the subject as a one size fits all. If this is the case, should robots be left to navigate their own evolutionary path alongside humans? This way, maybe they will have to grow into this world rather than being placed into this world. It calls into question how society should implement robots alongside humans. At present, if many of our fears are around the aesthetics of robots and their potential function of taking over human jobs, then we need to question the place that robots will have among us – and arguably, challenge globalist views and approaches. Maybe a more localised, culturally and geographically focused approach would be more suitable and effective in the long term when placing robots alongside humans. This could certainly be beneficial from a tourism and hospitality perspective. It would allow, as proposed here, for the desire for future tourists (humans and potentially robots) to travel to new places to learn from other humans and robots. If human interaction when travelling offers us new opportunities for learning about ourselves and others, then would society not want robots that, like us, are unique to their environment? If robots are to be valued, then their developers need to provide them with the opportunity to be valued. Thus, cultural relativity should be a fundamental element in robots and part of their evolution. Why would I travel to another part of the world to interact with a robot that I can have at home. Thus, if robots are to play an active role in the motivations and desires of tourists, then they will need to offer something beyond the familiarities of what tourists can experience at home.

References

Allain, R. (2015) The robotification of society is coming. https://www.wired.com/2015/01/robotification-society-coming/ (accessed 6 January 2021).
Appel, M., Marker, C. and Mara, M. (2019) Otakuism and the appeal of sex robots. *Frontiers in Psychology* 10 (569), 1–11.
Bahishti, A.A. (2017) Humanoid robots and human society. *Advanced Journal of Social Science* 1 (1), 60–63.
Brown, G. (2020) *Difference between humanoid robot and android*. http://www.differencebetween.net/technology/difference-between-humanoid-robot-and-android/#ixzz6il7aCShE (accessed 6 January 2021).
Caraian, S., Kirchner, N. and Colborne-Veel, P. (2015) Moderating a robot's ability to influence people through its level of sociocontextual interactivity. In *Proceedings of the Tenth Annual ACM/IEEE International Conference on Human–Robot Interaction* (pp. 149–156). ACM.
Carnevale, A. (2015) Robots, disability and good human life. *Disability Studies Quarterly* 35 (1).
Chatila, R. (2018) Inclusion of humanoid robots in human society: Ethical issues. In P. Vadakkepat and A. Goswami (eds) *Humanoid Robotics: A Reference* (pp. 2665–2674). Springer.
Choi, C.Q. (2008) Not tonight, dear, I have to reboot. https://www.scientificamerican.com/article/not-tonight-dear-i-have-to-reboot/ (accessed 6 January 2021).
Connell, J. (2006) Medical tourism: Sea, sun, sand and… surgery. *Tourism Management* 27 (6), 420–434.
Cook, P.S. (2010) Constructions and experiences of authenticity in medical tourism: The performances of places, spaces, practices, objects and bodies. *Tourist Studies* 10 (2), 135–153.
Cresswell, K., Cunningham-Burley, S. and Sheikh, A. (2018) Health care robotics: Qualitative exploration of key challenges and future directions. *Journal of Medical Internet Research* 20 (7), 1–11.
Dang, S.S. (2019) Artificial intelligence in humanoid robots. https://www.forbes.com/sites/cognitiveworld/2019/02/25/artificial-intelligence-in-humanoid-robots/?sh=475e435924c7 (accessed 6 January 2021).

Devlin, K. (2015) In defence of sex machines: Why trying to ban sex robots is wrong. The conversation. https://theconversation.com/in-defence-of-sex-machines-why-trying-to-ban-sex-robots-is-wrong-47641 (accessed 6. January 2021).

Duncan, D.E. (2019) *Talking to Robots: A Brief Guide to Our Human–Robot Futures*. Robinson.

Durães, D., Bajo, J. and Novais, P. (2018) Characterize a human–robot interaction: Robot personal assistance. In A. Costa, V. Julian and P. Novais (eds) *Personal Assistants: Emerging Computational Technologies. Intelligent Systems Reference Library* (vol. 132, pp. 135–147). Springer.

Eichenberg, C. and Küsel, C. (2018) Roboter in der Psychotherapie: Intelligente artifizielle Systeme. https://www.aerzteblatt.de/pdf.asp?id=199391 (accessed 8 January 2021).

Fetscherin, M. and Stephano, R.-E. (2016) The medical tourism index: Scale development and validation. *Tourism Management* 52, 539–556.

Fiske, A., Henningsen, P. and Buyx, A. (2019) Your robot therapist will see you now: Ethical implications of embodied artificial intelligence in psychiatry, psychology, and psychotherapy. *Journal of Medical Internet Research* 21 (5), 1–12, e13216.

Friedman, G. (2009) *The Next 100 Years: A Forecast for the 21st Century*. Anchor Books, Random House.

Ghosh, T., and Mandal, S. (2018) Medical tourism experience:Conceptualization, scale development, and validation. *Journal of Travel Research* 58 (8), 1288–1301.

Global Wellness Institute (2018) *Global Wellness Economy Monitor*. Global Wellness Institute.

Goldapple, L. (2020) The sex robots are coming… https://atlasofthefuture.org/the-sex-robots-are-coming/ (accessed 8 January 2021).

Huijnen, C.A.G.J., Lexis, M.A.S., Jansens, R. and de Witte, L.P. (2017) How to implement robots in interventions for children with autism? A co-creation study involving people with autism, parents and professionals. *Journal of Autism and Developmental Disorders* 7 (10), 3079–3096.

Hutton, M. (2019) Sex tourism is changing as travellers seek pleasure in adult-only resorts. Can Asia keep up? https://www.scmp.com/magazines/post-magazine/travel/article/3033000/sex-tourism-changing-travellers-seek-pleasure-adult (accessed 8 January 2021).

IAMAT (2017) Implications of sexual tourism. https://www.iamat.org/blog/implications-of-sexual-tourism/#:~:text=Sexual%20tourism%20is%20a%20very,over%20%20%24 20%20billion%20in%20revenue (accessed 8 January 2021).

Ivanov, S. and Webster, C. (2018) Adoption of robots, artificial intelligence and service automation by travel, tourism and hospitality companies: A cost–benefit analysis. In V. Marinov, M. Vodenska, M. Assenova and E. Dogramadjieva (eds) *Traditions and Innovations in Contemporary Tourism* (pp. 190–203). Cambridge Scholars Publishing.

Ivanov, S., Gretzel, U., Berezina, K., Sigala, M. and Webster, C. (2019) Progress on robotics in hospitality and tourism: A review of the literature. *Journal of Hospitality and Tourism Technology* 10 (4), 489–521.

Jones, G. (1985) *Divine Endurance*. Allen & Unwin.

Koskinen, V. (2019) Spa tourism as a part of ageing well. *International Journal of Spa and wellness,* 2(1), 18-34.

Larson, R.W. (2002) Globalization, societal change, and new technologies: What they mean for the future of adolescence. *Journal of Research on Adolescence* 12 (1), 1–30.

Levy, D. (2008) *Love and Sex with Robots: The Evolution of Human–Robot Relationships*. Harper Perennial.

Liddell, H.G., Scott, R., Jones, H.S. and McKenzie, R. (1996) *A Greek–English Lexicon* (9th edn). Clarendon Press.

Lindgren, M. and Bandhold, H. (2009) *Scenario Planning: The Link between Future and Strategy*. Palgrave Macmillan.

Lorber, J. and Moore, L.J. (2010) *Gendered Bodies: Feminist Perspectives* (2nd edn). Oxford University Press.

Lumidolls (2021) Barcelona brothel. https://lumidolls.com/en/hotel-barcelona (accessed 8 January 2021).

McDonald, C. (2015) The good, the bad and the robot: Experts are trying to make machines be 'moral'. https://alumni.berkeley.edu/california-magazine/just-in/2015-06-08/good-bad-and-robot-experts-are-trying-make-machines-be-moral (accessed 6 January 2021).

Merrifield, R. (2013) *Integrating smart robots into society*. https://horizon-magazine.eu/article/integrating-smart-robots-society.html (accessed 6 January 2021).

Mori, M. (1970/2005) *The Uncanny Valley*. (trans. K.F. MacDorman and T. Minato). *Energy* 7, 33–35.

Mueller, H. and Kaufmann, E.L. (2001) Wellness tourism: Market analysis of a special health tourism segment and implications for the hotel industry. *Journal of Vacation Marketing* 7 (1), 5–17.

Open Access Government (2019) What if we had robot therapists instead of humans? https://www.openaccessgovernment.org/robot-therapists/65433/ (accessed 8 January 2021).

Pearson, I. (2015) The future of sex report: The rise of the robosexuals. http://graphics.bondara.com/Future_sex_report.pdf (accessed 11 December 2020).

Richardson, K. (2016) Sex robot matters: Slavery, the prostituted, and the rights of machines. *IEEE Technology and Society Magazine* 35 (2), 46–53.

Rogers, S.A. (2021) *Almost human: 15 frighteningly realistic robots & androids*. https://weburbanist.com/2014/06/30/almost-human-15-frighteningly-realistic-robots-androids/ (accessed 6 January 2021).

Saunderson, S.P. and Nejat, G. (2021) Persuasive robots should avoid authority: The effects of formal and real authority on persuasion in human–robot interaction. *Science Robotics* 6 (58), 1–11.

Schwab, K. (2016) The Fourth Industrial Revolution: What it means, how to respond. https://www.weforum.org/agenda/2016/01/the-fourth-industrial-revolution-what-it-means-and-how-to-respond/ (accessed 8 January 2021).

Schwartz, P. (1991) *The Art of the Long View: Planning for the Future in an Uncertain World*. Doubleday.

Selterman, D. (2015) Explainer: What is sexual fluidity? https://theconversation.com/explainer-what-is-sexual-fluidity-33120 (accessed 6 January 2021).

Smith, M. and Kelly, C. (2006) Wellness tourism. *Tourism Recreation Research* 31 (1), 1–4.

Smith, M.K. and Puczkó, L. (2014) *Health, Tourism and Hospitality: Spas, Wellness and Medical Travel*. Routledge.

Specktor, B. (2018) Meet Erica, Japan's next robot news anchor. https://www.livescience.com/61575-erica-robot-replace-japanese-news-anchor.html (accessed 6 January 2021).

Stafford, A. (2016) Android clone v human: Will you be able to tell the difference at work? https://www.theguardian.com/sustainable-business/2016/nov/03/android-clone-v-human-will-you-be-able-to-tell-the-difference-at-work (accessed 6 January 2021).

Sullins, J.P. (2012) Robots, love, and sex: The ethics of building a love machine. *IEEE Transactions on Affective Computing* 3 (4), 398–409.

Tangermann, V. (2017) Six life-like robots that prove the future of human evolution is synthetic. https://futurism.com/the-most-life-life-robots-ever-created (accessed 6 January 2021).

Tatsumi, T. (2006) *Full Metal Apache: Transactions between Cyberpunk Japan and Avant-Pop America*. Duke University Press.

The Medical Futurist (2018) The future of sex and sexuality. https://medicalfuturist.com/the-future-of-sex-and-sexuality/ (accessed 11 December 2020).

The Oxford English Dictionary (2015) Futurology. https://www.lexico.com/definition/futurology (accessed 4 January 2022).

Watson, A. (2016) It was an ancient form of sex tourism. https://www.bbc.com/culture/article/20161017-it-was-an-ancient-form-of-sex-tourism (accessed 8 January 2021).

World Health Organisation (WHO) (2021) Gender and health. https://www.who.int/health-topics/gender#tab=tab_1 (accessed 6 January 2021).

Wright, D.W.W. (2016) Hunting humans: A future for tourism in 2200. *Futures* 78, 34–46.

Wright, D.W.W. (2019) Sport hunting and tourism in the twenty-second century: Humans as the ultimate trophy. *Foresight* 21 (3), 419–442

Yeoman, I. and Mars, M. (2012) Robots, men and sex tourism. *Futures* 44 (4), 365–371.

10 The Sustainability of Tourism in Robonomics

Craig Webster, Fernando J. Garrigos-Simon
and Yeamduan Narangajavana-Kaosiri

10.1 Introduction

In recent decades there has been a concern for the preservation of the natural environment which has developed into the concept of sustainability. The concept is broader than a superficial concern with protections for the natural environment and includes ecological and social concerns. The Brundtland Report (WCED, 1987) and the United Nation's (UN) 17 sustainable development goals (SDGs) are illustrations that global institutions and the social expectations of the world's political elite are interested in forwarding this notion in order to protect the environment and create an interaction between human activity and the sustainability of the ecosystem upon which the human economy is dependent. Tourism sustainability, as the study and practice of sustainability in the tourism sector, can be defined as 'those actions and developments in the tourism arena that meet the needs of present tourists and host societies without having a negative impact on the environment, ecology, society, landscape, culture, and patrimony, and without compromising the prosperity and well-being of future generation' (Garrigos-Simon et al., 2018).

Sustainability, as a concept and ideology, maintains that humans are the core of the economy but live within a physical environment that nurtures humans. Beyond that, sustainability also includes social elements, essentially leading to a perspective that humans should function in an ecosystem in which the economy serves humanity and the environment is protected in ways that ensure that there is a long-term trajectory that enables human flourishing. Here, we focus on the relationship between the robonomic system and how the characteristics of such a system will influence the environmental, social and economic dimensions of the sustainability of tourism. We first look into the major characteristics of robonomics and how these characteristics will support a sustainable future. Then, we look into how these characteristics will become incorporated into a robonomic tourism ecosystem, in which people, planet and profits are supported and nurtured by the advent and the incorporation of automation technologies. Finally, we reflect on the challenges to attain a sustainable robonomic tourism ecosystem and the nature and logic of such a system to serve humans in the not-so-distant future.

10.2 Robonomics and Sustainability

The robonomic society will have a direct influence on the three components of sustainability typically referred to as people, planet and profits, from Elkington's (1998) conceptualisation of the triple bottom line (see Table 10.1). We look at how each of these characteristics of robonomics will impact on companies' triple bottom line.

10.2.1 High level of automation

The most obvious and clear impact of the robonomic system on sustainability will be the high level of automation. Automation technologies will go beyond industrial production or performing menial tasks such as delivering food or cleaning floors but will carry out various tasks that require cognition and analysis and emotional awareness (Huang *et al.*, 2019). Automation technologies will be incorporated into all industries

Table 10.1 Robonomics' characteristics and influences on sustainability

Characteristics of robonomics	Major influence on sustainability
(1) High level of automation	• Increased efficiency in the use of land, labour and capital in industry and society • General reduction in negative externality, including waste materials • Creation of unemployment and growth of inequalities
(2) Artificial autonomous agents serve as economic agents	• Increase decision-making efficiencies • Avoidance of human-based emotionality of decision-making • Decisions might be divorced from the ethical concerns of humans
(3) Few humans work	• Decrease in energy and time associated with the transportation of employees to a workplace • Decrease in waste in industry based on human error • Decreased costs associated with human labour • Avoidance of labour stoppages/strikes • Displacement of humans from the workforce and stresses on social institutions
(4) Disconnection between employment and incomes	• Decreased seasonality in human consumption • Increased stability and predictability of the economic system • Undermining of current financial, political and economic institutions
(5) Sources of competitive advantage	• Efficient use of humans as creative entities • Exacerbating inequalities between those with creative capabilities and those without them
(6) Überveillance	• Ability to identify inefficiencies and destructive behaviours • Increased efficiency in production specific to individual human tastes, wants and needs • Undermining of privacy and threats to individual rights

so that goods and services will be produced with few or no human agents. Artificial intelligence (AI) technologies will ensure both energy efficiency and a reduction in waste. There is already some indication that the use of automation technologies has a positive impact on the physical environment, although the relationship may not be entirely linear (Yang et al., 2023). With increased technological capacity to provide efficient use of the factors of production, inputs into the economy such as natural resources, human labour and capital can be used in ways that are most efficient (Ivanov, 2017) and create the least negative externalities (such as pollution) possible.

While technological advances are shown to increase productivity and profits (Frey, 2019), such advances will require a great deal of energy (and there are questions about the sustainability of various sources), may include materials that are toxic in nature and may involve processes that seem socially unpleasant or environmentally polluting. Current widely used lithium batteries are toxic (Liu et al., 2021; Nedjalkov et al., 2016) as are plastics (Verma et al., 2016) and some extraction industries, such as the mining of Coltan, are associated with very negative social consequences (Ayres, 2012; Bleischwitz et al., 2012; Mantz, 2008). However, these are some of the major technological and social challenges that technological advances may be able to solve in order to guarantee that the outcomes are sustainable. For example, mining is increasingly automated (Paredes & Fleming-Muñoz, 2021), which has the positive social externality of reducing the exploitation of workforces for the extraction of needed minerals from the ground. There is also reason to believe that automation may work in ways to undermine long-term concerns with sustainability apart from narrow environmental concerns. For example, automation will create large-scale unemployment, may undermine some aspects of the quality of work and it may even exacerbate gender equality, apart from other social externalities that will remain unknown unknowns.

In terms of the triple bottom line, this means that automation will positively impact on the quality of life of all people, whether consumer or employee (Webster & Ivanov, 2023). Humans will be spared from doing dirty, dull, dangerous and repetitive tasks that make the workplace unpleasant and lead to medical and psychological stresses that are not good for humans. In terms of the planet, automation technologies will reach new heights of efficiency in the use of raw materials and energy and decrease the amount of waste being generated, since technology will be used to ensure that goods and services are not just of high quality and inexpensive but also produced creating the fewest negative externalities possible. At the same time, industry should benefit, as the reduction in the use of resources to manage humans and the more efficient use of land, labour and capital will increase profitability.

10.2.2 Artificial autonomous agents serve as economic agents

Artificial autonomous agents will play a key role as economic agents, making better and more efficient decisions and avoiding mistakes associated with human error or emotionality. While humans may be able to make judgments that are based on sound assumptions and knowledge, humans do not have the computing capacity of modern and future automation technologies (Giddings, 2008). In addition, humans sometimes act on hunches or emotions to make decisions (Lerner et al., 2015). Although emotionality may assist in making good decisions, they are not consistent and may lead to bad decisions with poor consequences (Lerner et al., 2015). Autonomous agents will be so prolific that they will be recognised as legal entities (in the way that corporations today are legal entities) and own property, which may eventually entail extending them some sort of rights (Gunkel, 2018). While there is a concern that autonomous agents may not have the ethics of a human and may not make choices that favour the outcomes that humans may favour, such ethics can be embedded into their programming and may be mandated by political authorities. For example, while autonomous agents may choose to purchase raw materials for an industrial enterprise because they are cheaper than more earth-friendly alternatives, the environmental costs of products can be factored into the algorithms of the autonomous agent to ensure that dangerous materials are not purchased.

People should benefit from autonomous agents making decisions because decisions made in society and the economy will be done in ways that are better informed and with greater considerations for the externalities than a simple human could, since such technologies can process far more data than a human (Giddings, 2008). Certainly, consumers will benefit from better prices when autonomous agents make decisions regarding industrial production and consumer needs, as such agents will promote the maximisation of efficiency and the quality of goods and services. The planet will benefit from such a scheme, as autonomous agents will demand the efficient use of resources and the avoidance of wasteful processes. Finally, companies should profit, as they will see massive increases in productivity while autonomous agents work to ensure the lowest costs from suppliers.

10.2.3 Few humans work

With fewer humans working, there should be a massive reduction in the costs of human error and human consumption associated with human labour. Workers will not use their time and energy transporting themselves to a workplace so that there will be material savings on energy production that is now associated with rush hour and the very inefficient externality of rush hour. While people may travel more for leisure purposes, such travel can be smoothed out and managed to avoid the inefficient movement of people, as what typically takes place in a rush hour scenario in major cities, where time, energy and space are used in a very inefficient manner. In addition, businesses will not have to dedicate space to human employees, such as break rooms, or budget for other expenses associated with training and socially engaging their workforces. Also, there will be a reduction in the energy used in workplaces for the comfort of humans (e.g. in terms of temperature and air quality) since the workplace environment in many places will be designed for automation technologies and not the narrow range of human sensitivities. A great deal of waste will be reduced in the workplace, as autonomous agents will make efficient and more informed decisions compared to humans. Finally, labour stoppages/strikes will not be effective nor practical, since automation levels will make the impact of labour union action futile.

There will be concerns as to what to do with humans who do not work, but it is likely that social innovations such as universal basic income (UBI), a scheme with some noted risks (Fouksman & Klein, 2019) will ensure that the social and economic system continues to function. There will also likely be problems with keeping people engaged in society in ways that promote a sense of purpose for individuals, so that volunteering opportunities, make-work projects and innovative leisure activities will have to be developed to promote a sense of belonging. The displacement of humans from the workforce will stress the existing institutions that regulate society so new institutions and methods for engaging humans in productive (or at least not destructive) activities will have to be developed. At any rate, the general movement away from large families around the world (Webster, 2021; Webster & Ivanov, 2020) will lead to a world with a smaller population (Coleman & Rowthorn, 2011) and facilitate the management of idle humans.

In terms of the triple bottom line for businesses, people will benefit from this as humans will experience a higher quality of life due to being relieved of workplace stresses. The planet will benefit, since resources will not be needed to transport people to and from work and supply workers with comfortable work environments. Finally, there will be changes in profits with automation encouraging productivity and avoiding inefficiencies due to human error, and an investment in resources dedicated to recruiting, training, retraining and otherwise making employees happy and comfortable in their workplaces. While total revenues may drop due to changes in human numbers and consumption patterns, decreased investments in human capital should increase the overall profitability of enterprises.

10.2.4 Disconnection between employment and incomes

Currently, the disconnection between employment and income is something mostly enjoyed by the very rich, the very poor or the retired.

However, in a robonomic society, this will be the predominant way in which people function, since little human labour will be needed to make the economy function in a meaningful and effective way. Many humans will not need to work and will receive UBI, meaning that they will not have to work nor schedule their lives around work nor certain seasonality issues. Freed from the scheduling imposed by an employer or a profession's needs and freed from the need to work specific hours on specific days for income, humans will be able to consume in ways that avoid bottlenecks of consumption. At the system level, the economic system may be much more stable, since there is an expectation that economic cycles of boom and bust may be mitigated by the steady and predictable basis of UBI. In total, the economy will be expected to be stable and predictable. The risk is that UBI may not be a sustainable basis on which to build an entire economy, so the successful social engineering of a sustainable UBI as a basis for the functioning of the economy is critical. There will be significant turbulence in the transition from the current salary-based economy since new institutions will have to be created or modified to ensure that a sustainable social and political order can be developed for a robonomic economic system to function.

In terms of the triple bottom line, people will not have the stress of performance-based incomes, some of which may be out of the control of the employee. Essentially, the bulk of humanity will be living as if they were retired with a decent and livable pension, choosing to work, if desired, and spending free time on leisure or self-betterment. The planet will benefit as many of the environmental costs associated with work such as the use of energy for commuting to and from work and the use of resources associated with workforces, such as uniforms and materials dedicated to employees, will be seriously diminished. Companies will profit by having a more stable market for the purchase of goods and services that will not be as dependent on the fluctuations of the market, instead being based on an almost continuous system of Keynesian investment in society, albeit bypassing many make-work projects.

10.2.5 Sources of competitive advantage

Sources of competitive advantage will also work in ways to promote sustainability in a robonomic world, since human-generated creativity will be used to limit waste, use resources more efficiently and create new possibilities for the economy. Competitive advantage will be decreasingly influenced by the cost of labour but will be based on the human's ability to envision things that have market value, something already recognised as a major contributor to the creation of value in our economy today (Wijaya & Suasih, 2020). The values of the entire robonomic system are based on the efficient use of energy and resources, so there will be a natural evolution of the ethic of the system towards sustainability, giving no marketing advantage to claims of sustainability. A concern is that since the basis of competitive advantage is creative abilities, many humans will lack such capabilities, so they will be largely unemployable and excluded from the productive side of the economy.

A negative outcome of this system is that an elite of decision-makers will decide what values are embedded in the entire system and conflicts may arise at times between environmental sustainability and social concerns now conceptualised as sustainability. For example, there may be a situation in which the use of human labour may be environmentally sustainable and efficient but unacceptable socially, if some specific task cannot be fully automated and efficient. In addition, it is possible that the autonomous agents could favour environmental protections above the human or determine that humans are damaging to the environment and undermine human well-being in subtle ways. For example, algorithms may favour inexpensive and earth-friendly food components that are less healthy for humans while being cheaper and better for the physical environment.

People will benefit from using the human mind in the creation of competitive advantage, as humans will be freed from working on menial tasks and used for what humans truly excel at – the creation of new things and ideas. The planet should benefit from such a shift to focus on human creativity since humans will create new ideas but automation technologies will implement the concepts in ways that are designed to be more sustainable for the natural environment. Finally, companies' profitability will expand, since humans will be unshackled from the limitations of the consideration of the engineering of final outcomes and will be able to concentrate on the visualisation of new ideas and products.

10.2.6 Überveillance

While überveillance (Michael & Michael, 2013), the notion of high levels of surveillance of humans, may seem to be an unpleasant concept, it will be a key characteristic of the robonomic society that will work in many ways to benefit sustainability. Most importantly, a great deal of data will be gathered and used to promote waste reduction, as knowledge of human consumption patterns will be made known through the collection and analysis of big data (Barham, 2017). Additionally, surveillance will ensure that the behaviours of humans or autonomous agents that are inefficient or destructive can be identified and dealt with in an effective, efficient and timely way. The major risk of such a system is sliding into an Orwellian world in which humans are objects of a totalitarian control system that puts immense pressures on human liberty (Boggs, 2016). The risk of moving from a rights-based liberal democratic framework to a utilitarian socially engineered and invasive state is real.

In terms of the triple bottom line, companies will see that criminality, including theft and vandalism, can be minimised, as surveillance tools will guarantee high levels of protection for individuals and private property. Surveillance also allows for big data to be collected in ways to ensure efficiency in the production of goods and services, since there will be substantial knowledge of the consumption and tastes of individuals. In that way, the planet is protected by ensuring the minimisation of waste in the use of inputs and energy to provide goods and services to customers. Finally, companies can expect gains in profitability, since there will be increased knowledge of consumer wants and needs, along with increased security of private property.

10.2.7 Robonomics and sustainability in gestalt

The overview of the entire system and its relationship to sustainability is that it will create a number of economic and ecological benefits, largely through massive increases in efficiency and reductions in waste. Such increases in the productivity of the system and reductions in waste will ultimately mean an increase in the quality and quantity of production. The massive increase in the economic potential may eliminate poverty and hunger and promote the health of the global population with clean water and affordable clean energy. However, it will also entail a change in the political and social order that may undermine the specific non-economic goals of the SDGs, leading to negative outcomes for humanity and our current understanding of the SDGs, unless intelligent institutional safeguards are put in place. For example, it is unclear if automation technologies will increase undesirable forms of discrimination (Seyitoğlu & Ivanov, 2023), something that could undermine the advancement of gender equality. In addition, it is likely that robonomics will encourage greater inequality, with the vast majority of humanity not being able to take part in the new economy, largely dependent on new welfare state institutions, and sometimes being relegated (counterintuitively) to the workplace tasks that are not yet able to be automated. Finally, an Orwellian world in which the state has access to vast amounts of data and surveillance of the population threatens to be a police state and this threatens our current conceptualisation of rights and justice that are enshrined in the SDGs.

10.3 Robonomics and Sustainability in the Tourism Ecosystem

In the robonomic world, the tourism ecosystem will be highly sustainable for a myriad of reasons. Firstly, the high levels of automation and the massive reduction of humans involved in the production of tourism-related products will lead to high-quality and low-cost hospitality and tourism services. While there may be a bifurcation of the market with the wealthy having the advantage of access to high-touch, rather than high-tech services (Ivanov, 2019), the bulk of humanity will have access to hospitality and tourism services that are run in the most efficient way and with the least waste of energy, human labour and materials. Consumers, for example, in restaurants and tourism accommodations will have little or no interaction with front-of-house and back-of-house human employees, since there are likely to be few or no humans actually performing these activities. While there may be a remaining market that demands and can pay for high-touch services, much of the market will be dependent on UBI and acclimated to interacting with automation technologies for the services that they desire. In addition, the technologies will engineer the smoothing out of tourism flows so that overcrowding and the inefficiencies and environmental impacts associated with it can be avoided.

Secondly, as artificial autonomous agents increasingly make decisions and own property, they will create an environment in which all services can be supplied at the lowest possible price, which is part of the efficiency that the robonomic system promises (Ivanov, 2017). Because of this, tourists and hospitality customers can be reassured that the food and drink that they order include inputs that are cost-effective, benefiting both the consumer and the service provider. Additionally, the quantity of data that autonomous agents will have available to them and the system of überveillance will work in conjunction to market to consumers in an efficient and effective way, avoid waste and protect tourists and customers in many different ways. A restaurant customer will be offered foods that they prefer, and the back-of-house will have the predictive mechanisms beforehand to minimise waste in the ordering of products for the inputs into menus.

Thirdly, since few humans will work and their spending will be independent of their employment status, a great deal of seasonality in the practice of tourism can be avoided, so the high season can be smoothed out. The ability of consumers to be freed from work schedules and probably also the schedules of schooling used for children's education will mean that the calendar of humans and their leisure will be less restrictive. While summers may remain the time when people prefer to holiday near beaches, autonomous agents can influence consumption with incentivised prices to discourage overtourism in some high-demand destinations. The autonomous agents' manipulation of the market prices will be used to ensure that the environmental strains of overtourism are mitigated and consumers will use their automation technologies to optimise their wealth to identify leisure times and costs that will maximise their leisure for the money invested. In this way, autonomous agents can work to ensure that aeroplanes minimise empty seats and avoid overbooking, to minimise losses. While perishability will remain a major feature of tourism and hospitality services, the robonomic system will ensure that empty aeroplane seats, empty rooms in hotels and food waste in restaurants are minimised by engineering prices and predictive analytics technologies.

Fourthly, humans will work in ways to add creativity to enterprises, inventing new products and services and imagining new ways to deliver new experiences for customers of tourism and hospitality services. In a way, this is the efficient use of humans, since a human strength is creative ability and it will lead to new and innovative forms of leisure and tourism, albeit supported by automation technologies, advanced analytical tools and big data. New forms of tourism and leisure that have not yet been invented or thought of because of the technological limitations in the current era will come into being, including leisure using such technological concepts as the Neuralink (Pisarchik *et al.*, 2019). The merger of computers and the human brain may permit not just the enhancement of human cognitive abilities but may also offer new opportunities for leisure, such as embedded memories. An embedded memory of a pleasant holiday may be a cost-effective and sustainable alternative to a physical holiday, require little energy and natural resources and, in the end, increase people's quality of life.

Finally, überveillance sounds unpleasant to those of us in the current day who have been socialised with an expectation of constitutional rights, liberties and privacy. However, there are some advantages. Mostly, the existence of massive amounts of data offers major competitive advantages to companies with access to it (Barham, 2017), but when such data becomes commonplace and is available to many autonomous agents, there will be some advantages with regard to the provision of services that are desired. The data on customers may be so detailed that autonomous agents will be able to provide goods and services with minimal waste, allowing for minimum expenditure of energy and material goods, supporting a sustainable tourism economy. Autonomous agents will have enough data to supply customers with specifically tailored services to maximise the consumer's enjoyment. For example, if the data on a particular customer in a restaurant is available, menu recommendations may only offer foods that the customer likes and exclude foods not acceptable to the customer (such as excluding non-Kosher foods to pious Jewish customers or those foods to which the customer is allergic). Such a system would lead to reduced food waste and a more efficient consumer experience. Additionally, it would be expected that such surveillance of the population will give a great sense of security to most tourists and customers, as autonomous agents should be able to predict, identify and prevent the destructive behaviours of humans. While überveillance sounds Orwellian, humans may be happy to surrender their privacy for better material outcomes and increased security in hospitality and tourism settings.

10.4 The Road to Sustainable Robonomic Tourism

On the way to a sustainable robonomic tourism ecosystem, a few challenges will have to be overcome and some transitional turbulence. First, automation technologies will have to be much more advanced than what is now available. While there has been a movement from robots simply doing routine manual tasks in industrial facilities towards automation technologies that do much more complicated tasks and collaborate with humans (Decker *et al.*, 2017), much more progress needs to be made. There will have to be a massive increase in the technological capabilities of automation technologies so that service failures are few and far between and require little human interaction to correct the failures. Only when the technological capabilities are so great that failures are few and far between will there be a strong consumer preference for service automation. However, there may remain consumers who prefer the authenticity of a human service (Seyitoğlu, 2021), even when technologies are effective and attractive due to cost savings.

Second, a major technological issue remains with the provision of energy sources that ensure the environmentally sustainable production and storage of energy. Freeing the economy from its dependence on petroleum (both for energy and as a raw material for production) will be a major issue and failure to do so may lead to a very unpleasant Mad Max type of future scenario for humanity (Webster, 2019). Alternative energy sources are also associated with sustainability problems. For example, the technology used to make solar power currently involves many pollutants and heavy metals (Jayapradha & Barik, 2023) so their production and disposal includes managing toxic and dangerous materials. In terms of energy storage, the lithium battery, the type of battery commonly used in many devices today, is toxic and unsafe (Liu *et al.*, 2021; Nedjalkov *et al.*, 2016), although efforts are being made to find safer and less noxious alternatives (Choi & Aurbach, 2016; Orikasa *et al.*, 2014). Lithium-ion batteries not only contain components that are unsavoury but when they burn (something they are prone to do) they emit toxic emissions (Larsson *et al.*, 2017). So, while we may have the technology to make energy from the sun and wind, the processes to capture the energy, the disposal of the outdated technologies used to capture the energy and the environmentally safe storage of the energy from sustainable sources are a major impediment to a sustainable future in which the robonomic tourism ecosystem will exist. To this end, there are opportunities for technological breakthroughs that will transform the economy and society. Sustainable energy sources and storage capabilities must be developed in order for a sustainable robonomic economy to exist.

Third, current robots are largely made from plastics. Plastics are a recurring issue with regard to environmental concerns (Verma *et al.*, 2016), and there is a question as to whether they can fit into an environmentally sustainable future (Zaman & Newman, 2021). New technologies will have to create materials to make robots and other automation technologies that are not damaging to the environment and the humans working in their environment. While plastics may be a cheap, effective, generally safe and lightweight material, they are likely too contaminating to be part of a long-term and sustainable future. Technological breakthroughs that create processes for earth-friendly materials that are non-toxic to humans will have to be made to ensure a truly sustainable future.

Fourth, we will need to have new social institutions to deal with a world in which humans are less prevalent in the supply of goods and services; indeed, social change is one of the critical components of the emerging economy's success (Skene, 2019). We need to cope with how überveillance, changing the tax base and shifting to the majority of people being dependent on UBI will work in conjunction with the new basis of the economy. While these factors may not play a critical and direct role in terms of environmental issues, these issues do play a critical role in terms of the social components of sustainability so that the political economy can support industry and the production of goods and services for humans.

What we know is that the quality of automation technologies will continue to improve and this should lead to a robonomic economy in which tourism and hospitality will play a key role. Tourism and hospitality will play an important role in the attainment even of the short-term goals of the SDGs (Ivanov et al., 2023) and will remain a major part of the human economy, often used as a tool for sustainable economic and social development, environmental protection (Sætra, 2022) and political stability. Once the major technological and social impediments to a more automated economy are overcome, the tourism ecosystem will function in ways that are sustainable and good for the planet, people and profits of companies. Future tourists will enjoy an enhanced quality of life because of robonomics (Webster & Ivanov, 2023) and partake in a robonomic tourism ecosystem that will be sustainable and available to all.

References

Ayres, C.J. (2012) The international trade in conflict minerals: Coltan. *Critical Perspectives on International Business* 8 (2), 178–193. https://doi.org/10.1108/17422041211230730

Barham, H. (2017) Achieving competitive advantage through big data: A literature review. In *2017 Portland International Conference on Management of Engineering and Technology (PICMET)* (pp. 1–7). IEEE. https://doi.org/10.23919/PICMET.2017.8125459

Bleischwitz, R., Dittrich, M. and Pierdicca, C. (2012) Coltan from Central Africa, international trade and implications for any certification. *Resources Policy* 37 (1), 19–29. https://doi.org/10.1016/j.resourpol.2011.12.008

Boggs, C. (2016) Technological rationality and the post-Orwellian society. *Glimpse* 17, 10–19. https://doi.org/10.5840/glimpse2016172

Choi, J.W. and Aurbach, D. (2016) Promise and reality of post-lithium-ion batteries with high energy densities. *Nature Reviews Materials* 1 (4), 1–16. https://doi.org/10.1038/natrevmats.2016.13

Coleman, D. and Rowthorn, R. (2011) Who's afraid of population decline? A critical examination of its consequences. *Population and Development Review* 37, 217–248. https://doi.org/10.1111/j.1728-4457.2011.00385.x

Decker, M., Fischer, M. and Ott, I. (2017) Service robotics and human labor: A first technology assessment of substitution and cooperation. *Robotics and Autonomous Systems* 87, 348–354. https://doi.org/10.1016/j.robot.2016.09.017

Elkington, J. (1998) Accounting for the triple bottom line. *Measuring Business Excellence* 2 (3), 18–22. https://doi.org/10.1108/eb025539

Fouksman, E. and Klein, E. (2019) Radical transformation or technological intervention? Two paths for universal basic income. *World Development* 122, 492–500. https://doi.org/10.1016/j.worlddev.2019.06.013

Frey, C.B. (2019) *The Technology Trap: Capital Labor, and Power in the Age of Automation*. Princeton University Press.

Garrigos-Simon, F.J., Narangajavana-Kaosiri, Y. and Lengua-Lengua, I. (2018) Tourism and sustainability: A bibliometric and visualization analysis. *Sustainability* 10 (6), 1976. https://doi.org/10.3390/su10061976

Giddings, G. (2008) Humans versus computers: Differences in their ability to absorb and process information for business decision purposes — and the implications for the future. *Business Information Review* 25 (1), 32–39. https://doi.org/10.1177/0266382107088211

Gunkel, D.J. (2018) *Robot Rights*. MIT Press.

Huang, M.H., Rust, R. and Maksimovic, V. (2019) The feeling economy: Managing in the next generation of artificial intelligence (AI). *California Management Review* 61 (4), 43–65. https://doi.org/10.1177/0008125619863436

Ivanov, S. (2017) Robonomics: Principles, benefits, challenges, solutions. *Yearbook of Varna University of Management* 10, 283–293.

Ivanov, S. (2019) Ultimate transformation: How will automation technologies disrupt the travel, tourism and hospitality industries? *Zeitschrift für Tourismuswissenschaft* 11 (1), 25–43. https://ssrn.com/abstract=3335811

Ivanov, S., Duglio, S. and Beltramo, R. (2023) Robots in tourism and sustainable development goals: Tourism Agenda 2030 perspective article. *Tourism Review* 78 (2), 352–360. https://doi.org/10.1108/TR-08-2022-0404

Jayapradha, P. and Barik, D. (2023) A review of solar photovoltaic power utilizations in India and impacts of segregation and safe disposal of toxic components from retired solar panels. *International Journal of Energy Research* 2023. https://doi.org/10.1155/2023/3196734

Larsson, F., Andersson, P., Blomqvist, P. and Mellander, B.E. (2017) Toxic fluoride gas emissions from lithium-ion battery fires. *Scientific Reports* 7 (1), 10018. https://doi.org/10.1038/s41598-017-09784-z

Lerner, J.S., Li, Y., Valdesolo, P. and Kassam, K.S. (2015) Emotion and decision making. *Annual Review of Psychology* 66, 799–823. https://doi.org/10.1146/annurev-psych-010213-115043

Liu, F., Wang, T., Liu, X. and Fan, L.Z. (2021) Challenges and recent progress on key materials for rechargeable magnesium batteries. *Advanced Energy Materials* 11 (2), 2000787. https://doi.org/10.1002/aenm.202000787

Mantz, J.W. (2008) Improvisational economies: Coltan production in the eastern Congo. *Social Anthropology/Anthropologie Sociale* 16 (1), 34–50. https://doi.org/10.1111/j.1469-8676.2008.00035.x

Michael, M.G. and Michael, K. (ed.) (2013) *Uberveillance and the Social Implications of Microchip Implants: Emerging Technologies*. IGI Global.

Nedjalkov, A., Meyer, J., Köhring, M., Doering, A., Angelmahr, M., Dahle, S., Sander, A., Fischer, A. and Schade, W. (2016) Toxic gas emissions from damaged lithium ion batteries: Analysis and safety enhancement solution. *Batteries* 2 (1), 5. https://doi.org/10.3390/batteries2010005

Orikasa, Y., Masese, T., Koyama, Y., Mori, T., Hattori, M., Yamamoto, K. ... Uchimoto, Y. (2014) High energy density rechargeable magnesium battery using earth-abundant and non-toxic elements. *Scientific Reports* 4 (1), 5622. https://doi.org/10.1038/srep05622

Paredes, D. and Fleming-Muñoz, D. (2021) Automation and robotics in mining: Jobs, income and inequality implications. *The Extractive Industries and Society* 8 (1), 189–193. https://doi.org/10.1016/j.exis.2021.01.004

Pisarchik, A.N., Maksimenko, V.A. and Hramov, A.E. (2019) From novel technology to novel applications: Comment on 'An integrated brain–machine interface platform with thousands of channels' by Elon Musk and Neuralink. *Journal of Medical Internet Research* 21 (10), e16356. https://doi.org/10.2196/16356

Seyitoğlu, F. (2021) Automation vs authenticity in services. *ROBONOMICS: The Journal of the Automated Economy* 2, 20. Retrieved from https://journal.robonomics.science/index.php/rj/article/view/20

Seyitoğlu, F. and Ivanov, S. (2023) Service robots and perceived discrimination in tourism and hospitality. *Tourism Management* 96, 104710.

Skene, K.R. (2019) *Artificial Intelligence and the Environmental Crisis: Can Technology Really Save the World?* Routledge.

Sætra, H.S. (2022) *AI for the Sustainable Development Goals* (1st edn). CRC Press. https://doi.org/10.1201/9781003193180

Verma, R., Vinoda, K.S., Papireddy, M. and Gowda, A.N.S. (2016) Toxic pollutants from plastic waste: A review. *Procedia Environmental Sciences* 35, 701–708. https://doi.org/10.1016/j.proenv.2016.07.069

WCED (1987) *Our Common Future*. Oxford University Press.

Webster, C. (2019) Halfway there: The transition from 1968 to 2068 in tourism and hospitality. *Zeitschrift für Tourismuswissenschaft* 11 (1), 5–23. https://doi.org/10.1515/tw-2019-0002

Webster, C. (2021) Demography as a driver of robonomics. *ROBONOMICS: The Journal of the Automated Economy* 1, 12. https://journal.robonomics.science/index.php/rj/article/view/12

Webster, C. and Ivanov, S. (2020) Demographic change as a driver for tourism automation. *Journal of Tourism Futures* 6 (3), 263–270. https://doi.org/10.1108/JTF-10-2019-0109

Webster, C. and Ivanov, S. (2023) Robots, artificial intelligence and service automation in tourism and quality of life. In M. Uysal and M.J. Sirgy (eds) *Handbook of Tourism and Quality-of-Life Research II: Enhancing the Lives of Tourists, Residents of Host Communities and Service Providers* (pp. 533–544). Springer International Publishing. https://doi.org/10.1007/978-3-031-31513-8_36

Wijaya, P.Y. and Suasih, N.N.R. (2020) The effect of knowledge management on competitive advantage and business performance: A study of silver craft SMEs. *Entrepreneurial Business and Economics Review* 8 (4), 105–121. https://doi.org/10.15678/EBER.2020.080406

Yang, X., Luan, F., Zhang, J. and Zhang, Z. (2023) Testing for quadratic impact of industrial robots on environmental performance and reaction to green technology and environmental cost. *Environmental Science and Pollution Research* 30, 92782–92800. https://doi-org.proxy.bsu.edu/10.1007/s11356-023-28864-4

Zaman, A. and Newman, P. (2021) Plastics: Are they part of the zero-waste agenda or the toxic-waste agenda? *Sustainable Earth* 4, 1–16. https://doi.org/10.1186/s42055-021-00043-8

11 Future Tourism in a Robonomic World: An AI-Generated Chapter

Stanislav Ivanov and Craig Webster

11.1 Introduction

In 2068, artificial intelligence (AI) will be pervasive. The previous chapters of this book highlighted our human visions of tourism in a robonomic society and pointed out that AI, robots and automation technologies will be behind the reception desk, behind the booking website, will prepare our morning coffee, will clean the room, will drive our vehicle, etc. These visions reflect our own human limitations and, probably, subconscious desires about possible tourism futures. Moreover, they all implicitly assume that we, humans, are the sole creators of our future. But this is not necessarily true. Recent advances in generative AI (Carvalho & Ivanov, 2024; Dogru *et al.*, 2023; Dwivedi *et al.*, 2023) allow us to investigate how AI perceives the future because it will likely also be a major actor in shaping the future.

This chapter presents an interview with OpenAI's ChatGPT (https://chat.openai.com/) about the future of tourism in a robot-based economy similar to previous studies (Fusté-Forné & Orea-Giner, 2023; Soliman & Al Balushi, 2023; Soliman *et al.*, 2023; Tarbit *et al.*, 2023). The conversation with ChatGPT took place between 15 May and 25 August 2023, utilising its paid GPT-4 version. The prompts introduced ChatGPT to the concept of robonomics as an economic system, asked about the future of the main tourism characteristic activities (passenger transport, accommodation, food and beverage, visitor attractions), explored the travel motives of future tourists, the drivers of corporate competitiveness on tourism suppliers in the future, the role of robots, AI and automation in tourism, and the skills of future tourism and hospitality employees. The interview with ChatGPT investigated the future major source markets and destinations, and the impacts of climate change on tourism. It forecasted the future of tourism industries, main destinations and source markets for countries in Europe, Asia, Africa and the Americas. Finally, the conversation contemplated the possibilities of having robots as tourists and giving robots some rights.

The text in the next section provides an excerpt of the whole conversation that is available at https://chat.openai.com/share/5bcdd610-5b19-4dd3-8c0e-cd5a12b280a2. The order of the prompts and answers has been amended for the sake of presentation. Section 11.3 provides a reflection on ChatGPT's answers and concludes the chapter.

11.2. Interview with ChatGPT

11.2.1 Future of economy

Q1: Hi, let's have a longer discussion on the future tourism. First, let us define 'robonomics' - it is an automated economy where most products and services are produced and delivered by robots, AI and automation technologies. What do you think tourism might look like in the future when the economy is largely automated?

A1: In a future where the economy is largely automated and driven by robonomics, tourism is likely to be transformed in a variety of ways. Here are some possible changes we might see in the tourism industry:

1. **Personalized experiences:** AI and big data analytics will enable a higher level of personalization in travel experiences. This means that each traveler can have a tailor-made itinerary based on their preferences, interests, and past behavior. Virtual assistants will help tourists make informed decisions about their trips, providing personalized recommendations for attractions, accommodations, and transportation.
2. **Virtual and augmented reality:** VR and AR technologies will play a significant role in the future of tourism, allowing people to explore destinations before even stepping foot in them. Virtual tours of popular attractions and landmarks, as well as augmented reality experiences overlaying information on real-world environments, will enrich the tourist experience and help them plan their trips more effectively.
3. **Seamless travel:** Automation and AI will streamline the entire travel process, making it more efficient and enjoyable. From airport check-ins to hotel bookings and guided tours, robots and AI-powered systems will facilitate seamless interactions, minimizing wait times and reducing human error.
4. **Eco-friendly and sustainable travel:** As robonomics leads to increased efficiency, the tourism industry will be better equipped to address environmental concerns. Sustainable travel options, like electric or solar-powered vehicles, and eco-friendly accommodations will be more accessible, and AI-driven systems will optimize resource usage to minimize waste.
5. **Space tourism:** With the advancements in technology and automation, space tourism will become more accessible and affordable.

Companies like SpaceX and Blue Origin are already working on commercial space travel, and with the support of automation, we might see space tourism becoming a reality for more people.
6. **Safety and security:** AI and automation will enhance safety and security measures in the tourism industry. Facial recognition technology, automated luggage scanning, and AI-driven threat detection systems will contribute to safer and more secure travel experiences.
7. **New job opportunities:** While automation will undoubtedly displace certain jobs in the tourism sector, it will also create new opportunities. Tourism professionals will need to adapt their skillsets to work alongside AI and robotic systems, focusing on more creative, strategic, and empathetic roles to enhance the human touch in the tourism experience.

In summary, the future of tourism in a robonomics-driven world is likely to be characterized by greater personalization, enhanced travel experiences, increased efficiency, a stronger focus on sustainability, and new opportunities in space tourism and job creation.

11.2.2 Future of transport

Q2: How will travel look like in 2068? What transportation vehicles will people use? Will there be some unconventional vehicles that do not exist yet? What about their speed, size, comfort and sustainability? How will travel costs change?

A2: While it is impossible to predict the exact developments in transportation by 2068, we can speculate on some possible trends and innovations that may shape travel during that time. Here are some potential characteristics of transportation in 2068:

1. **High-speed travel:** As technology continues to advance, we may see the development of ultra-fast transportation options such as the Hyperloop or supersonic planes, drastically reducing travel times between cities and even continents. These modes of transportation will likely prioritize efficiency, speed, and passenger comfort.
2. **Autonomous vehicles:** By 2068, autonomous vehicles will likely be the norm, both on the ground and in the air. Self-driving cars and taxis will make ground travel more efficient, while autonomous drones and flying cars could revolutionize urban mobility and air travel.
3. **Sustainable transportation:** There will likely be a continued emphasis on developing eco-friendly transportation options, such as electric or hydrogen-powered vehicles, to reduce the carbon footprint of travel. We may also see advanced biofuels playing a more significant role in air travel and marine transportation.

4. **Space travel:** As space tourism continues to develop, we may see more accessible and affordable options for trips to Earth's orbit or even lunar bases. Spaceplanes or reusable rockets might become more commonplace, allowing for both scientific and leisure trips to space.
5. **Interconnected transport networks:** Integrated transportation systems that seamlessly connect different modes of travel (e.g., trains, buses, and autonomous taxis) could make door-to-door journeys more efficient and convenient. Smart city infrastructures will help optimize traffic flow and reduce congestion.
6. **Personal mobility devices:** Advances in personal mobility technologies may give rise to new types of individual transportation devices that don't exist yet. These could include advanced electric bikes, scooters, hoverboards, or even personal exoskeletons that enable faster and more efficient movement within urban environments.
7. **Travel costs:** With increased efficiency and reduced reliance on fossil fuels, travel costs may decrease for some modes of transportation. However, factors such as increased demand for travel, advancements in technology, and economic factors will also impact prices, making it difficult to predict the overall trend in travel costs.

In summary, transportation in 2068 could be characterized by high-speed, autonomous, and sustainable options that prioritize efficiency, comfort, and eco-friendliness. Advances in technology and infrastructure could lead to the emergence of new transportation vehicles, while factors like reduced fossil fuel dependency and increased efficiency may impact travel costs. However, it is essential to keep in mind that these predictions are speculative, and the actual development of transportation in 2068 will depend on numerous factors.

11.2.3 Future of accommodation

Q3: Thank you. Let us now move to the accommodation in the same year (2068). What accommodation establishments would be prevailing? What services will tourists use in them? Will there be any special design characteristics for the accommodations? Any unusual types?

A3: In 2068, the accommodation landscape might be considerably different from what we know today, driven by technological advancements, changing traveler preferences, and environmental concerns. Here are some potential trends and innovations we might see in accommodations during that time:

1. **Smart hotels:** Hotels in 2068 will likely be equipped with advanced technology that improves guest experiences. Rooms might be fully automated and customizable, with features such as voice-activated

controls, adaptable lighting, temperature regulation, and even adjustable layouts. Virtual concierges could assist guests with their needs, while AI-driven systems will optimize resource consumption for energy efficiency.
2. **Space-saving designs:** As urban spaces become increasingly crowded, accommodation establishments may adopt innovative design solutions to maximize space. This could include modular rooms that can be adapted for different purposes, transformable furniture, and multi-functional spaces. These designs will likely prioritize efficiency, flexibility, and guest comfort.
3. **Eco-friendly accommodations:** Sustainable practices will become more critical in the hospitality industry, and hotels will likely be designed with a strong focus on minimizing environmental impact. This could involve using renewable energy sources, incorporating green roofs or vertical gardens, employing water-saving technologies, and utilizing sustainable materials for construction.
4. **Unique accommodations:** As travelers continue to seek out unique experiences, we might see the rise of unconventional accommodation options that cater to various interests and preferences. Examples include underwater hotels, treehouses, floating lodges, and even space hotels for those engaging in space tourism.
5. **Blended accommodations:** The concept of co-living and co-working spaces may continue to evolve, leading to more accommodations that blend work, leisure, and living. These spaces will be designed to foster collaboration, creativity, and community, catering to digital nomads, remote workers, and long-term travelers.
6. **Immersive experiences:** Hotels and other accommodations may place a stronger emphasis on providing immersive experiences, enabling guests to connect more deeply with local cultures and environments. This could involve offering authentic, locally-sourced cuisine, organizing cultural events, or even designing accommodations that reflect traditional architectural styles.
7. **Personalized services:** AI and data analytics will enable hotels to offer highly personalized experiences to guests. They may have access to preferences and interests, which will allow them to provide tailored recommendations for activities, dining, and local attractions. The guest experience will be highly customized, making each stay more enjoyable and unique.

In summary, accommodations in 2068 are likely to prioritize technological advancements, sustainability, innovative design, and unique experiences. We might see the emergence of unconventional accommodation types that cater to diverse traveler preferences, as well as more personalized services enabled by AI and data analytics. However, these predictions are speculative, and the actual development of accommodations in 2068 will depend on various factors.

11.2.4 Future of F&B

Q4: Thank you. What about Food and Beverage (F&B). What F&B outlets will be there? What foods and drinks might be popular? Any new foods? How will hospitality providers create unique and memorable gastronomic experiences for their customers?

A4: The Food and Beverage (F&B) industry in 2068 is likely to evolve significantly, driven by factors such as technological advancements, environmental concerns, and changing consumer preferences. Here are some potential trends and innovations we might see in F&B outlets:

1. **Alternative proteins:** As concerns about the environmental impact and ethical considerations of traditional animal agriculture grow, alternative protein sources like plant-based meat substitutes, lab-grown meat, and insect-based protein may become more popular. These alternatives can provide the taste and texture of traditional meat products while reducing environmental impact and addressing food security issues.
2. **Personalized nutrition:** Advances in nutrition science, AI, and data analytics could enable highly personalized meal planning and food choices that cater to individual health needs and preferences. F&B outlets may offer customized menus that take into account dietary requirements, allergies, or personal health goals, ensuring a more satisfying and health-conscious dining experience.
3. **Sustainable practices:** Environmental consciousness will likely play a more significant role in the F&B industry. We might see more outlets focusing on reducing food waste, utilizing locally-sourced ingredients, adopting circular economy principles, and using eco-friendly packaging. Consumers will increasingly prioritize sustainability when choosing dining options.
4. **Culinary innovations:** As technology continues to advance, we may see innovative cooking techniques and new food products. Molecular gastronomy, 3D food printing, and precision fermentation could create novel taste experiences, textures, and combinations that challenge traditional culinary boundaries.
5. **Immersive dining experiences:** F&B outlets might focus on providing unique and memorable gastronomic experiences, incorporating elements of storytelling, interactivity, and sensory immersion. This could involve using virtual reality, augmented reality, or projection mapping to create immersive dining environments, or hosting themed events that transport guests to different times, cultures, or imaginary worlds.

6. **Automation and robotics:** The use of AI and robotics in the F&B industry may become more prevalent, with tasks such as food preparation, cooking, and service being performed by machines. This could lead to increased efficiency, consistency, and hygiene in food production, while also allowing human staff to focus on more creative and customer-centric roles.
7. **Health-focused food and beverages:** The growing awareness of health and wellness may lead to an increased demand for functional foods and beverages that offer specific health benefits. This could include probiotics, fortified products, or plant-based offerings that cater to health-conscious consumers.

In summary, F&B outlets in 2068 are likely to prioritize sustainability, personalized nutrition, innovative culinary techniques, and unique dining experiences. We might see a shift toward alternative proteins, eco-friendly practices, and health-focused products, as well as an increased role for technology and automation in food preparation and service. However, it is essential to keep in mind that these predictions are speculative, and the actual development of F&B in 2068 will depend on numerous factors.

11.2.5 Future of visitor attractions

Q5: Thank you! Let's focus on the visitor attractions (museums, galleries, theme parks, amusement parks, etc.) and additional services and tours (e.g. spa and wellness, safari, photo safari, paragliding, rafting, etc.). How will these look like in 2068? Any new and extraordinary attractions/services/tours that can be offered to tourists?

A5: In 2068, visitor attractions and additional services and tours are likely to be shaped by advancements in technology, a focus on sustainability, and the desire for unique and immersive experiences. Here are some potential developments we might see in these areas:

1. **Immersive attractions:** Attractions such as museums, galleries, and theme parks will likely focus on providing more immersive experiences. Virtual reality, augmented reality, and holography could be widely used to create interactive exhibits, simulations, and storytelling experiences that engage visitors on a deeper level.
2. **Gamification and interactivity:** Visitor attractions may increasingly incorporate elements of gamification, turning exhibits or attractions into interactive experiences where visitors can participate, learn, and compete with others. This approach can make learning more enjoyable and engaging for people of all ages.
3. **Personalized experiences:** Attractions and tours may offer personalized experiences based on visitor preferences and interests, using

AI and data analytics to tailor content and recommendations. This could include custom itineraries for museum visits or guided tours that focus on specific themes or subjects of interest.
4. **Sustainable practices:** As sustainability becomes more important, attractions and tour operators may adopt eco-friendly practices and promote environmental awareness. This could involve using renewable energy sources, minimizing waste, and providing educational experiences about the local environment and conservation efforts.
5. **Virtual and remote experiences:** Advancements in technology may make it possible for people to explore attractions or participate in tours without physically being present. Virtual tours, live streams, and remote-controlled experiences could become more prevalent, allowing people to travel and explore from the comfort of their homes.
6. **Extreme and adventure tourism:** As people continue to seek new experiences and challenges, extreme and adventure tourism may grow in popularity. This could involve activities like space tourism, deep-sea exploration, or extreme sports that push the boundaries of human capabilities and experiences.
7. **Health and wellness tourism:** With a growing focus on health and well-being, we may see a rise in services and experiences centered around wellness, mindfulness, and holistic health. This could include retreats, spa and wellness centers, and therapeutic experiences designed to promote physical, mental, and emotional well-being.
8. **Cultural and educational tourism:** There may be an increased emphasis on providing authentic cultural and educational experiences that allow visitors to learn about and engage with local communities, traditions, and history. This could involve hands-on workshops, cultural exchanges, or participatory events that foster understanding and appreciation of different cultures.

In summary, visitor attractions and additional services and tours in 2068 may prioritize immersive experiences, personalization, sustainability, and a focus on health, wellness, and cultural understanding. New and extraordinary attractions and services could emerge as technology advances and people seek out unique experiences that push the boundaries of what's possible. However, these predictions are speculative, and the actual development of these offerings in 2068 will depend on various factors.

11.2.6 Motives of future tourists

Q6: Thank you. What will be the travel motives of future tourists? Why would they undertake a trip? Which factors will drive their purchase decisions?

A6: The travel motives of future tourists may evolve as societal values, preferences, and technological advancements change over time. Here are some possible travel motives and factors that could drive purchase decisions in the future:

1. **Unique and authentic experiences:** As people increasingly seek meaningful experiences, travelers may prioritize destinations and activities that offer something unique and authentic, allowing them to connect with local cultures, history, and nature on a deeper level.
2. **Personal growth and education:** Future tourists may be driven by the desire to learn and grow personally. Travel could be seen as an opportunity to acquire new skills, expand cultural horizons, and gain a deeper understanding of the world and its diverse people.
3. **Health and wellness:** With a growing focus on health and well-being, future tourists may be more inclined to undertake trips centered around wellness, relaxation, and rejuvenation. This could involve spa retreats, spiritual journeys, or outdoor adventures that promote physical and mental health.
4. **Sustainability and eco-tourism:** Environmental concerns and the desire to minimize one's ecological footprint may drive future tourists to choose eco-friendly destinations, accommodations, and experiences. Sustainable tourism practices and responsible travel will likely become more important factors in decision-making.
5. **Adventure and extreme experiences:** The pursuit of adrenaline, excitement, and unique experiences may lead future tourists to choose more adventurous and extreme activities or destinations, such as space tourism, deep-sea exploration, or remote wilderness adventures.
6. **Digital detox and disconnection:** As our lives become more intertwined with technology, some future tourists may be driven to seek out destinations and experiences that offer a break from constant connectivity. Trips centered around digital detox, disconnection, and reconnection with nature or oneself may become more popular.
7. **Social influence and personalization:** The role of social media and peer recommendations may continue to grow, influencing travel decisions and preferences. Future tourists may rely more heavily on personalized recommendations from friends, family, or influencers when making travel choices.
8. **Customization and convenience:** Advances in technology, AI, and data analytics will enable higher levels of personalization and convenience in travel planning. Future tourists may prioritize travel options that offer a seamless, tailor-made experience, with technology simplifying the planning and decision-making process.

In summary, the travel motives of future tourists are likely to include unique experiences, personal growth, health and wellness, sustainability, adventure, digital detox, social influence, and customization. These factors, along with technological advancements, will likely shape their preferences and purchase decisions as they seek out travel experiences that align with their values and desires.

11.2.7 Future of corporate competitiveness

Q7: What will be the drivers of a tourism/hospitality company's competitiveness in 2068?

A7: In 2068, the competitiveness of tourism and hospitality companies may be influenced by a variety of factors. Here are some possible drivers of competitiveness for these businesses in the future:

1. **Technological innovation:** Companies that embrace and leverage cutting-edge technology will likely have a competitive advantage. Innovations in AI, automation, virtual reality, and data analytics can help improve customer experiences, operational efficiency, and offerings.
2. **Sustainability:** Environmental consciousness and sustainable practices will be increasingly important in the tourism and hospitality industry. Companies that prioritize eco-friendly operations, promote responsible tourism, and minimize their environmental impact will likely be more competitive and attractive to consumers.
3. **Unique and authentic experiences:** Providing memorable and authentic experiences will be crucial for attracting customers. Companies that can offer innovative, immersive, and personalized experiences will have an edge over their competitors.
4. **Personalization and customization:** The ability to offer tailor-made experiences and services to meet individual customer preferences and needs will be increasingly important. Using data analytics and AI to understand customer preferences and deliver personalized offerings can help businesses differentiate themselves.
5. **Excellent customer service:** Outstanding customer service will continue to be a significant driver of competitiveness in the tourism and hospitality industry. Companies that invest in employee training, foster a customer-centric culture, and utilize technology to enhance service will likely gain a competitive advantage.
6. **Health and wellness focus:** With the growing importance of health and well-being, companies that incorporate wellness offerings into their services and experiences can differentiate themselves and appeal to health-conscious travelers.
7. **Strategic partnerships and collaboration:** Forming strategic alliances with other businesses, such as transportation providers, technology

companies, or local attractions, can help tourism and hospitality companies expand their offerings, improve customer experiences, and increase their market reach.
8. **Agility and adaptability:** The ability to quickly adapt to changing market conditions, consumer preferences, and industry trends will be crucial for businesses to remain competitive. Companies that can innovate, embrace change, and adjust their strategies to stay ahead of the curve will be better positioned for long-term success.
9. **Strong online presence and digital marketing:** An effective online presence and digital marketing strategy will be essential for reaching potential customers and showcasing a company's offerings. Utilizing social media, online reviews, and content marketing will play a crucial role in attracting and engaging customers.
10. **Focus on safety and security:** Ensuring the safety and security of guests will be a significant factor in a company's competitiveness. Companies that prioritize and invest in safety measures, including cyber-security and physical security, will likely build trust with customers and gain a competitive edge.

In summary, the drivers of competitiveness for tourism and hospitality companies in 2068 may include technological innovation, sustainability, unique experiences, personalization, customer service, health and wellness, strategic partnerships, agility, online presence, and safety and security. Companies that can excel in these areas and adapt to the evolving market demands and consumer preferences will be well-positioned to succeed in the future.

11.2.8 The future role of robots, AI and automation in tourism

Q8: What will be the role of robots, artificial intelligence and automation technologies in tourism in 2068?

A8: In 2068, robots, artificial intelligence (AI), and automation technologies could play a significant role in the tourism industry, reshaping how various services and experiences are provided. Here are some potential ways these technologies could be utilized in tourism:

1. **Customer service:** AI-powered chatbots and virtual assistants could handle customer inquiries, reservations, and bookings, providing fast and efficient service. Robots may also assist in front desk operations, providing personalized customer service and support.
2. **Personalization:** AI and data analytics could enable highly personalized travel recommendations, itineraries, and experiences based on individual preferences, interests, and past behavior. This would allow the tourism industry to cater more effectively to the needs and desires of travelers.

3. **Accommodations:** Robots and automation technologies could take care of various tasks within hotels, such as room service, housekeeping, and maintenance, improving efficiency and quality of service. Smart hotels with automated systems may offer customizable room settings, lighting, and temperature controls, enhancing guest comfort and satisfaction.
4. **Food and beverage:** Automation could play a significant role in food preparation, cooking, and service, leading to increased efficiency, consistency, and hygiene. Robotic waiters or AI-driven systems could provide personalized food and drink recommendations based on customer preferences, dietary requirements, or health goals.
5. **Transportation:** Autonomous vehicles, including self-driving cars, buses, and shuttles, could streamline transportation for tourists, offering safer and more efficient travel options. AI could also optimize travel routes and schedules, reducing congestion and environmental impact.
6. **Visitor attractions and experiences:** AI, virtual reality, augmented reality, and robotics could be integrated into various tourist attractions and experiences, creating immersive and interactive environments. Virtual guides, holographic displays, and gamified experiences could engage and educate visitors in innovative ways.
7. **Travel planning and booking:** AI-driven platforms could simplify the travel planning process, aggregating and analyzing data from various sources to recommend the best travel options, accommodations, and experiences based on individual preferences and budgets.
8. **Safety and security:** AI and robotics could enhance safety and security measures in the tourism industry. Facial recognition technology, AI-powered surveillance systems, and robotic security patrols may help maintain a secure environment for travelers.
9. **Data-driven decision making:** AI and data analytics could provide valuable insights into customer behavior and preferences, enabling tourism businesses to make more informed decisions about their services, marketing strategies, and investments.
10. **Sustainability:** AI and automation technologies could contribute to more sustainable tourism practices by optimizing energy consumption, waste management, and transportation. This would help reduce the industry's environmental impact and promote eco-friendly tourism.

In summary, robots, AI, and automation technologies could play a pivotal role in reshaping the tourism industry in 2068, improving efficiency, personalization, sustainability, and the overall travel experience. These technologies have the potential to streamline various aspects of the industry, from customer service to transportation and visitor attractions, ultimately revolutionizing how people experience travel.

11.2.9 Skills of future employees in tourism and hospitality

Q9: In 2068, what types of education and skills will humans need to have to have successful careers in hospitality and tourism?

A9: In 2068, the landscape of education and skills required for successful careers in hospitality and tourism will likely have evolved significantly due to technological advancements, environmental concerns, sociocultural shifts, and other global trends. Here are some potential types of education and skills that could be vital:

1. **Technological Proficiency:**
   - **Digital literacy:** The ability to effectively use various digital tools, platforms and software will be a fundamental requirement.
   - **AI and Automation Management:** Understanding how to work alongside AI systems, and perhaps even some basics in programming or managing these systems.
   - **Virtual & Augmented Reality:** Designing and managing immersive experiences for guests may be integral to various hospitality sectors.
2. **Sustainability and Eco-consciousness:**
   - Understanding of sustainable practices in hospitality and how to implement eco-friendly solutions.
   - Expertise in sustainable tourism, promoting eco-friendly travel, and managing environmental impacts.
3. **Cultural Intelligence & Global Awareness:**
   - Being knowledgeable about various world cultures, understanding cultural sensitivities, and being able to offer tailored experiences to a diverse clientele.
   - Language skills, given the global nature of tourism.
4. **Emotional Intelligence & Human-centric Skills:**
   - Even with AI and automation, the hospitality sector will always be about people. Thus, interpersonal skills, empathy, and the ability to understand and manage one's own emotions, as well as those of others, will be crucial.
   - Conflict resolution and negotiation skills, especially in customer-facing roles.
5. **Adaptability & Continuous Learning:**
   - Given the rapid pace of change, professionals in 2068 will need to be adaptable and committed to lifelong learning.
   - Problem-solving and critical thinking to navigate unforeseen challenges.
6. **Entrepreneurship & Innovation:**
   - As tourists seek unique experiences, the ability to innovate and think entrepreneurially will be valuable.
   - Business acumen to understand market dynamics and manage operations efficiently.

7. **Specialized Knowledge & Skills:**
   - Depending on the niche within tourism (e.g., space tourism, deep-sea experiences, virtual world tours), professionals might need specialized training and qualifications.
8. **Ethics and Privacy:**
   - With the rise of technology and data collection, understanding ethical considerations, digital rights, and privacy issues will be crucial.
9. **Health and Well-being:**
   - Expertise in health, wellness, and possibly even medical tourism might be important given the increasing focus on holistic well-being.
10. **Safety and Security:**
    - Understanding both physical safety measures and cyber-security will be crucial, especially with the integration of more technology into the industry.
11. **Multidisciplinary Education:**
    - With the blurring of industry boundaries and the integration of tech, design, health, and more into tourism, having education that bridges multiple disciplines might be beneficial.

Traditional educational pathways (university degrees, diplomas) will likely coexist and possibly intermingle with alternative education platforms, like online courses, virtual reality learning experiences, and real-world apprenticeships. Furthermore, continuous upskilling and reskilling will probably be a norm given the evolving nature of the industry.

In essence, while technological skills will be vital, the unique human touch that defines the hospitality and tourism industry will remain indispensable. Balancing technology with interpersonal skills will be key to success.

11.2.10 Future major source markets

Q10: In 2068, which countries will be the ten major source countries for international tourism flows?

A10: Predicting specific international tourism flow for 2068 is challenging given the numerous variables at play, from economic growth and political stability to environmental conditions and technological advancements. However, based on current trends (up to 2021) and broader global predictions, we can make some educated guesses:

1. **China:** Given its large population and rapid economic growth over the past few decades, China is likely to remain a significant source of international tourists if this trend continues.
2. **India:** With a population set to surpass China's and an emerging middle class, India has the potential to become one of the most significant sources of international tourists.

3. **United States:** Historically a major source of international tourists due to its economic strength and size, the U.S. is likely to remain an influential player in global tourism.
4. **African Countries (e.g. Nigeria or South Africa):** Africa is predicted to experience significant population growth, and if accompanied by economic growth, countries like Nigeria, South Africa, or even Ethiopia might emerge as major sources of international tourists.
5. **Indonesia:** As the largest economy in Southeast Asia with a significant population, Indonesia could see a rise in outbound tourism if its economic trajectory continues upward.
6. **Brazil:** As the largest economy in South America, Brazil has the potential to be a significant source of international tourists, provided it sees stable economic growth and political stability.
7. **Vietnam:** With rapid economic growth and a young population, Vietnam might emerge as a strong player in international tourism over the next few decades.
8. **Russia:** As a vast country with abundant resources and a sizeable population, Russia might continue to be a significant source of international tourists if its economy remains stable.
9. **Middle Eastern Countries (e.g. Saudi Arabia, UAE):** With diversification efforts away from oil and towards sectors like tourism, entertainment, and technology, countries like Saudi Arabia or the UAE might drive outbound tourism as their populations seek international experiences.
10. **European countries (e.g. Germany, UK, or France):** Europe has traditionally been a strong source of international tourists. Depending on political and economic conditions, countries like Germany, the UK, or France might remain influential in global tourism, though they may see competition from emerging economies.

It's essential to remember that these predictions are speculative and can be influenced by unforeseen factors like geopolitical events, environmental challenges (like climate change), global pandemics, technological disruptions, and shifts in global economic power.

Furthermore, the concept of "tourism" might change by 2068. With the potential of virtual tourism, space tourism, and other yet-to-be-conceived travel forms, the very nature of what we consider "travel" might be very different from today.

11.2.11 Future tourist destinations

Q11: In 2068, which destinations will be the leading five destinations to which international tourists will travel?

A11: Predicting the top destinations in 2068 is highly speculative due to the myriad of factors that influence travel choices. Yet, building on existing trends, potential global developments, and a bit of foresight, we can make some educated guesses:

1. **Space Destinations:** By 2068, space tourism might not just be a luxury for the ultra-wealthy. Commercial space travel to the Moon, orbital hotels, or even Mars colonies could be the most sought-after experiences. Companies like SpaceX, Blue Origin and others are working towards these goals even now.
2. **Virtual and Augmented Realities:** Though not a 'destination' in the traditional sense, immersive VR and AR worlds could be a prominent 'travel' choice. People might choose to 'visit' historical periods, fictional worlds, or even reimagined versions of existing cities without leaving their homes.
3. **Underwater Cities or Resorts:** With advancements in technology, traveling to underwater accommodations in the deep sea could be an exotic experience. The allure of experiencing marine life up close might make destinations like the Great Barrier Reef or the Maldives (with its underwater villa) even more appealing if they adapt to the changing environmental conditions.
4. **Less-Explored Regions Turned Popular:** Places like Antarctica, Greenland, or remote islands in the Pacific and Indian oceans, currently seen as off-the-beaten-path, might become hotspots due to changing climate conditions and the human desire for novel experiences. Conversely, traditional tourist destinations suffering from over-tourism today might have evolved into sustainable tourism models that make them attractive again.
5. **Eco and Sustainable Destinations:** Given the increasing awareness of climate change and environmental issues, destinations that prioritize sustainability and eco-friendly practices might top the list. Countries like Costa Rica, New Zealand, or Bhutan, which emphasize sustainable tourism and conservation, might be leaders in this regard.

Factors that will heavily influence these destinations include:

- **Climate Change:** Rising sea levels, increased temperatures, and changing weather patterns might make some traditional destinations less appealing or even uninhabitable, while others might benefit from these changes.
- **Geopolitical Stability:** Countries that can maintain political stability, peace, and security will likely be more attractive to tourists.
- **Innovations in Travel:** If hyperloops, supersonic planes, or other rapid travel means become commonplace, it might drastically change the accessibility of certain destinations.
- **Cultural and Natural Preservation:** Destinations that can preserve their cultural heritage and natural wonders, especially in the face of global challenges, will continue to attract tourists.

Remember, these are speculative predictions, and the actual leading destinations in 2068 could be influenced by many unforeseen factors. However, the continuous evolution of the travel industry, combined with technological advancements, promises a future filled with exciting and diverse destinations.

11.2.12 Climate change

Q12: In 2068, how will climate change have made an impact upon the global tourism industry.

A12: By 2068, climate change is expected to have profound impacts on the global tourism industry. The implications of climate change are vast and multifaceted, encompassing environmental, socio-economic, and political dimensions. Here's a speculative look at some potential impacts:

1. **Shift in Traditional Tourist Destinations:**
   - **Beach Destinations:** Rising sea levels might affect numerous coastal regions and islands, reducing beach areas or even submerging some entirely. Destinations like the Maldives or certain Caribbean islands could see significant land losses. Coral reefs, critical attractions for many beach destinations, might also suffer due to ocean acidification and rising temperatures.
   - **Mountain Destinations:** Glacial retreat and reduced snowfall could change the landscape of popular mountain destinations, affecting winter sports and the overall allure of places known for their snowy peaks.
   - **New Destinations:** Some regions previously considered too cold or inhospitable might become new tourist hotspots as they become milder or unveil previously hidden attractions.
2. **Loss of Biodiversity:**
   - The loss of habitats and changing climate conditions might result in reduced biodiversity, impacting eco-tourism and safaris. Destinations known for specific species might no longer host those creatures. This could impact regions of Africa known for the "Big Five" or places like the Galapagos Islands.
3. **Increased Seasonality:**
   - Some regions might see shorter tourist seasons. For instance, ski resorts might have shorter winter seasons, and beach destinations might see reduced "pleasant" weather windows due to excessive heat or unpredictable weather patterns.
4. **Infrastructure Threats:**
   - Coastal infrastructure, from airports to hotels, might be at risk from rising sea levels and increased storm surges. More frequent and severe weather events can disrupt travel plans and pose challenges to infrastructure.

5. **Health and Safety Concerns:**
    - Increased temperatures and changing weather patterns can lead to health issues, from heat strokes to the spread of tropical diseases to regions where they weren't previously found.
    - Natural disasters, intensified by climate change, might deter tourists from certain destinations.
6. **Cultural Loss:**
    - Indigenous communities and their cultures, which often rely on specific environmental conditions, might be threatened, leading to a loss of cultural tourism opportunities.
7. **Migration and Geopolitical mpacts:**
    - Areas severely affected by climate change might see significant population migrations, leading to potential geopolitical tensions. Tourist destinations might change based on these new population centers or be impacted by regional conflicts resulting from resource scarcity.
8. **Sustainability as a Priority:**
    - Tourists might prioritize sustainable and eco-friendly destinations, leading to an industry shift towards green practices. Carbon offsetting, sustainable accommodations, and eco-tourism might become standard industry practices.
9. **Economic Shifts:**
    - Countries heavily reliant on tourism revenues from vulnerable destinations might experience economic downturns. Conversely, regions that adapt to or benefit from changing climate conditions might see economic benefits.
10. **Travel Behavior Changes:**
    - Tourists might choose to travel during different seasons, opt for closer-to-home vacations, or even favor virtual travel to reduce their carbon footprint.

The potential impacts of climate change on global tourism are significant and, in many cases, concerning. However, they also present opportunities for innovation, adaptation, and the prioritization of sustainable practices within the industry. The tourism sector, with its vast reach and economic clout, can also play a vital role in global efforts to mitigate the impacts of climate change.

11.2.13 Future of tourism in Europe

Q13: How would you characterize the hospitality and tourism industries in countries in Europe in 2068? Which countries will be most competitive in hospitality and tourism and why? Which will be their most important source markets?

A13: By 2068, Europe's deep-rooted history, culture, and attractions will likely remain strong magnets for tourists. However, the face of European tourism might evolve significantly, influenced by global trends, technological advancements, socio-political changes, and environmental factors.

Characteristics of the European Hospitality and Tourism Industries by 2068:

1. **Sustainable and Eco-Tourism:** Europe may pioneer sustainable tourism models with its cities and countries emphasizing green travel, carbon-neutral experiences, and conservation.
2. **Cultural and Historical Revival:** While Europe's historical and cultural sites are already major attractions, there could be a resurgence of interest in lesser-known histories, indigenous cultures, and rediscovered traditions.
3. **Technological Integration:** European cities might offer hyper-personalized travel experiences through AI, AR, and VR. Imagine virtual guides, real-time translations, or historical reenactments at archaeological sites.
4. **Health and Wellness Tourism:** Capitalizing on its natural landscapes and historical health retreats, Europe could become a hub for wellness travel, offering holistic experiences from Alpine retreats to Mediterranean wellness cruises.
5. **Space and Science Tourism:** As space travel becomes more accessible, countries involved in space exploration, like France and Germany, might offer space-related tourism experiences.

Leading Countries and Their Competitive Edges:

1. **France:** Its rich history, culture, and world-class cuisine will continue to be major draws. The country might also integrate high-tech experiences, from VR museum tours to AI-guided wine tastings.
2. **Germany:** With its mix of historical sites and modern cities, Germany could be at the forefront of technological tourism. Its strong engineering and tech sectors might birth unique travel experiences.
3. **Italy:** The allure of its history, art, and cuisine will remain timeless. Italy might also expand eco-tourism offerings in its diverse landscapes, from Sicilian beaches to Tuscan countryside.
4. **Scandinavian Countries (Norway, Sweden, Denmark, Finland):** These could lead in sustainable tourism, offering pristine natural experiences from Northern Lights viewing to forest retreats.
5. **Eastern European Countries (e.g. Poland, Hungary, Romania):** These nations, rich in tradition and history, might gain more attention for offering authentic experiences away from the traditionally crowded Western European hotspots.

Major Source Markets:

1. **Asia:** With the growing middle class and increasing connectivity, countries like China, India, and Southeast Asian nations could be significant sources of tourists for Europe.
2. **Africa:** As African economies grow and the middle class expands, more African tourists might explore Europe.
3. **Americas:** The US, Canada and Latin American countries will likely remain significant source markets.
4. **Within Europe:** Intra-European travel will continue to thrive, given the continent's interconnectedness and diversity.
5. **Middle East:** With evolving geopolitical dynamics and economic growth, countries like the UAE, Saudi Arabia, and others might contribute more significantly to European tourism.

Factors Influencing Competitiveness:

1. **Innovative Tourism Models:** Countries offering unique experiences, leveraging technology or niche cultural elements, will stand out.
2. **Sustainability:** With climate change at the forefront, nations practicing and promoting sustainable tourism will gain favor.
3. **Safety and Infrastructure:** Stability, safety, and modern infrastructure will continue to be key determinants.
4. **Accessibility:** Ease of travel, visa regulations, and connectivity will play crucial roles.

By 2068, Europe, with its incredible mosaic of cultures, histories, and landscapes, will have evolved its tourism offerings to align with global trends while staying rooted in its rich traditions. How individual countries adapt and innovate will determine their competitive edge in the global tourism landscape.

11.2.14 Future tourism in Africa

Q14: How would you characterize the hospitality and tourism industries in countries in Africa in 2068? Which countries will be most competitive in hospitality and tourism and why? Which will be their most important source markets?

A14: In 2068, Africa, with its rich tapestry of cultures, landscapes, and wildlife, could witness a tremendous evolution in its hospitality and tourism industries. Here's a speculative insight into what the future might hold:

Characteristics of the African Hospitality and Tourism Industries by 2068:

1. **Wildlife and Eco-Tourism:** Africa's unparalleled wildlife and natural landscapes will remain major draws. Efforts might be intensified to promote eco-tourism, focusing on sustainability and conservation.
2. **Cultural and Heritage Tourism:** The diverse cultures and histories across African countries could be more deeply explored, with immersive experiences delving into indigenous traditions, colonial histories, and modern African renaissance.
3. **Technological Integration**: Some African cities might emerge as tech hubs, offering tourists modern urban experiences, complete with AR and VR historical and cultural tours.
4. **Adventure and Sports Tourism:** From Saharan adventures to mountain trekking in countries like Tanzania and Kenya, Africa could expand its offerings in this segment.
5. **Health and Wellness:** Building on traditional African wellness practices, nations might develop unique wellness retreats.

Leading Countries and Their Competitive Edges:

1. **South Africa:** With its combination of wildlife, beaches, and vibrant cities like Cape Town and Johannesburg, South Africa could continue to be a major tourist destination. Its wine regions might also gain more global recognition.
2. **Kenya & Tanzania:** The Great Migration and iconic national parks will remain a draw. Coastal tourism along the Indian Ocean, combined with Mount Kilimanjaro treks, could provide diverse offerings.
3. **Morocco:** The blend of Berber, Arab, and French cultures, combined with its markets, landscapes, and historical sites, could continue to attract tourists.
4. **Egypt:** As a cradle of civilization, Egypt's ancient wonders, combined with Red Sea resorts, will remain major attractions.
5. **Ghana & Senegal:** With initiatives like "Year of Return," these countries might further develop heritage and cultural tourism, attracting the African diaspora and others interested in the transatlantic slave trade's history.

Major Source Markets:

1. **Asia:** As with Europe, Asia's growing middle class, particularly from China and India, could increasingly venture to Africa.
2. **Americas:** Beyond traditional tourists, there might be a rise in visitors from the African diaspora, exploring their roots and contributing to cultural exchanges.
3. **Europe:** The historical and economic ties between Europe and Africa will likely continue to drive European tourists to the continent.

4. **Within Africa:** As the continent's economic landscape transforms, intra-African tourism could see a significant boost.
5. **Middle East:** Proximity and developing economic ties might lead to increased tourism from Middle Eastern nations.

Factors Influencing Competitiveness:

1. **Infrastructure Development:** Countries investing in transportation, accommodation, and other infrastructural elements will have an edge.
2. **Political Stability and Safety:** Countries ensuring safety and political stability will be more attractive to international tourists.
3. **Conservation Efforts:** As global consciousness about the environment grows, nations prioritizing wildlife and environmental conservation will gain favor.
4. **Cultural Exchange Initiatives:** Programs encouraging cultural exchanges and deep dives into authentic African experiences will be a draw.
5. **Accessibility and Connectivity:** Improved air, rail, and road connectivity will be key determinants of tourism growth.

By 2068, Africa, with its unique offerings and potential for growth, could have carved out a more prominent space in the global tourism map. How individual countries harness their resources, combined with stability and sustainable practices, will determine their competitiveness.

11.2.15 Future tourism in Asia

Q15: How would you characterize the hospitality and tourism industries in countries in Asia in 2068? Which countries will be most competitive in hospitality and tourism and why? Which will be their most important source markets?

A15: Asia, with its incredibly diverse cultures, landscapes, and histories, is already a major hub for tourism. By 2068, given its economic growth and technological advancements, it's likely the continent will have further solidified its place on the global tourism map.

Characteristics of the Asian Hospitality and Tourism Industries by 2068:

1. **Technological Tourism:** Asian countries, being at the forefront of technological innovations, might offer tours and experiences centered around technology. Think futuristic cities, AI-driven cultural experiences, and VR-based historical tours.
2. **Eco and Sustainable Tourism:** With parts of Asia facing significant ecological challenges, there may be a pivot towards eco-tourism, emphasizing the protection of natural resources and landscapes.

3. **Spiritual and Wellness Tourism:** Building upon traditions of spirituality and wellness, countries may offer retreats focusing on meditation, yoga, traditional medicine, and holistic well-being.
4. **Culinary Tourism:** Asia's diverse and rich culinary traditions could be further highlighted, with countries offering immersive food tours, cooking classes, and farm-to-table experiences.
5. **Cultural Festivals and Events:** Asia's vast array of festivals, from India's Diwali to China's Spring Festival, could become major tourism draws.

Leading Countries and Their Competitive Edges:

1. **China:** As a global superpower, its blend of ancient culture and futuristic cities will continue to be a major attraction. Regions like Tibet and Yunnan could draw those seeking unique cultural experiences.
2. **India:** Its cultural diversity, historical sites, and natural beauty, from the Himalayas to the backwaters of Kerala, will remain significant draws. India might also emphasize spiritual tourism, focusing on yoga and Ayurveda.
3. **Japan:** A blend of the traditional (like Kyoto's temples) and the ultra-modern (Tokyo's tech-driven experiences) will keep Japan at the forefront. Japanese hospitality, or "Omotenashi", will remain a unique selling point.
4. **Southeast Asian Nations (Thailand, Vietnam, Indonesia, Philippines):** Their pristine beaches, rich histories, and diverse cultures will remain major attractions. Eco-resorts and sustainable tourism might become more prominent.
5. **South Korea:** Its contemporary culture, K-pop, dramas, and unique traditions, along with tech-driven experiences, might make South Korea a hotspot.

Major Source Markets:

1. **Within Asia:** The growing middle classes of countries like India, China, and Indonesia could lead to a surge in intra-Asian travel.
2. **Africa:** Given expected economic growth and a rising middle class in African countries, more Africans might venture to Asia for tourism.
3. **Americas:** Cultural and business exchanges, along with improved connectivity, could continue to bring tourists from North and South America.
4. **Europe:** European tourists, seeking diverse experiences and warmer climates, might continue flocking to Asia.
5. **Middle East:** Economic collaborations and improved connectivity might lead to increased tourist exchanges.

Factors Influencing Competitiveness:

1. **Sustainable Practices:** Countries that adopt sustainable tourism practices, preserving their natural and cultural heritages, will have an edge.
2. **Safety and Infrastructure:** Modern infrastructure, transportation systems, and a reputation for safety will be major determinants.
3. **Innovative Offerings:** Nations that offer unique, tech-driven or culturally immersive experiences will stand out.
4. **Ease of Travel:** Simplified visa regulations, good air connectivity, and integrated transportation systems will play a significant role.
5. **Preservation of Identity:** While adapting to global trends, countries that maintain and showcase their unique cultural identities will attract discerning travelers.

By 2068, Asia's blend of the ancient and the futuristic, coupled with its adaptability and innovation, could make it the epicenter of global tourism. How individual nations leverage their unique offerings while addressing challenges will determine their standing in the global tourism industry.

11.2.16 Future tourism in the Americas

Q16: How would you characterize the hospitality and tourism industries in countries in the Americas in 2068? Which countries will be most competitive in hospitality and tourism and why? Which will be their most important source markets?

A16: The Americas, encompassing North, Central, and South America, boast a rich tapestry of cultures, landscapes, and histories. By 2068, the continents could witness remarkable changes in their hospitality and tourism industries, influenced by technological advancements, shifting global priorities, and changing demographics.

Characteristics of the Hospitality and Tourism Industries in the Americas by 2068:

1. **Sustainable Tourism:** Given the rich biodiversity and fragile ecosystems in many parts of the Americas, eco-tourism and sustainable tourism practices might be paramount.
2. **Cultural and Heritage Tourism:** The vast indigenous cultures, colonial histories, and modern evolutions would provide a myriad of immersive experiences for tourists.
3. **Adventure Tourism:** From the Andes to the Rockies, the Amazon to the Arctic, adventure tourism could see even greater interest.
4. **Urban and Tech Tourism:** With some of the world's leading tech hubs in the Americas, cities could offer futuristic urban experiences, complemented by AR/VR enhancements.

5. **Wellness Retreats:** Given the rich traditional knowledge on well-being, especially in Central and South America, wellness tourism could flourish.

Leading Countries and Their Competitive Edges:

1. **United States:** Its diverse landscapes, iconic cities, and cultural impact, combined with technological prowess, will likely continue to make it a major tourism hub. Expect more eco-focused destinations and even space tourism facilitated by American companies.
2. **Canada:** Canada's vast natural beauty, from its national parks to the Northern Lights, could draw ecotourists. The country's multicultural cities might also offer unique cultural experiences.
3. **Mexico:** Beyond its beaches, Mexico's rich history, and vibrant culture will remain major draws. Eco-tourism and community-driven tourism initiatives might gain more prominence.
4. **Brazil**: The allure of the Amazon, its vibrant cities like Rio, and diverse cultural experiences could make Brazil a key destination.
5. **Argentina and Chile:** From Patagonian adventures to wine tours, these countries could offer a mix of natural beauty and cultural experiences.

Major Source Markets:

1. **Asia:** The growing affluence in Asia, particularly China, India, and Southeast Asian nations, could lead to increased tourism in the Americas.
2. **Within the Americas:** Domestic and intra-continental travel might see a rise, driven by improved connectivity and regional collaborations.
3. **Europe:** Europe's historical ties and interests in the Americas will continue to drive European tourists to the continent.
4. **Africa:** As African economies grow and global interconnectivity improves, expect a rise in African tourists.
5. **Middle East:** Economic collaborations and cultural exchanges might result in increased tourism from this region.

Factors Influencing Competitiveness:

1. **Sustainability:** Nations emphasizing sustainable practices and conservation will be preferred destinations.
2. **Technology Integration:** The infusion of tech into the tourism experience, from AR historical tours to AI-driven personal guides, will be a determinant.
3. **Safety and Political Stability:** Countries ensuring a safe environment for tourists and political stability will attract more visitors.

4. **Diverse Offerings:** Countries offering a range of experiences, from urban to rural, cultural to adventure, will have an edge.
5. **Connectivity and Accessibility:** Efficient transportation systems, both regional and international, will be crucial.

By 2068, the Americas, leveraging their diverse offerings and adapting to global trends, could provide some of the most enriching and varied tourism experiences in the world. The success of individual nations will hinge on their ability to innovate while preserving and showcasing their unique identities.

11.2.17 Robots as tourists

Q17: In 2068, the world should be quite different from now politically and economically. To what extent will robots be tourists, rather than employees? What facilities will be created to offer services to robot tourists?

A17: The concept of robots as tourists is a fascinating leap into the realms of speculative fiction and futurism. While it's hard to predict with certainty the specifics of 2068, the idea that robots or AI might "experience" tourism is predicated on several assumptions:

1. **Advanced AI Consciousness:** For robots to be tourists in the sense that humans are, they would need to have a level of consciousness that allows them to appreciate, experience, and seek out new experiences. As of the last update in 2021, no AI possesses consciousness, emotions, or subjective experiences.
2. **Value Proposition:** There must be a reason or value in sending robots as tourists. This could be for the collection of data, experiencing environments hazardous to humans, or as proxies for human tourists.
3. **Economic Model:** If robots are consumers of tourism, there must be an economic model that supports this. Who pays for the robot's experiences? Is it humans wanting a unique perspective? Or is it other AIs with economic interests?

Assuming a future where robots have some form of consciousness or value-driven reason to be tourists:
Facilities and Services for Robot Tourists:

1. **Data Exchange Centers:** Instead of traditional hotels or accommodation, robots might need centers where they can exchange, upload, or download data.
2. **Energy Stations:** Just as humans need food and water, robots would need places to recharge or refuel. These could be ubiquitous, especially in high-traffic tourist areas.

3. **Sensory Enhancement Facilities:** If robots are touring for sensory experiences, there might be locations where they can "upgrade" or "tune" their sensory inputs for optimal reception, whether it's visual, auditory, or even more advanced senses we haven't conceived of.
4. **Experience Simulators:** For AIs or robots that don't need to physically move to experience a place, virtual or simulated environments might be provided that give a detailed and rich "experience" of a destination.
5. **Informational Tours:** Robots might be interested in historical data, geology, environmental conditions, or other information-rich experiences. Tours might be tailored to provide vast amounts of data succinctly.
6. **Physical Interaction Zones:** If physical interaction is valuable, areas might be set up where robots can experience different terrains, elements, or interact with various objects, fauna, or flora in ways meaningful to their design.
7. **Communication Hubs:** Places where robots can interact with local AI systems, networks, or even other robot tourists to share experiences or information.
8. **Proxy Experience Centers:** If robots serve as proxies for human tourists, there might be centers where humans can "tune in" to their robot's experiences, guiding their journey or receiving sensory feedback.

The concept of robot tourism pushes the boundaries of current thinking around tourism, AI, and robotics. It forces us to consider the nature of experience, consciousness, and the very reasons we travel. Whether or not 2068 will see robot tourists in the way we imagine is uncertain, but pondering such possibilities stretches our understanding of both technology and our own human experience.

11.2.18 Robot rights

Q18: What are the characteristics of the robot and AI that will lead to the legal considerations that a robot and its software will be treated with the same responsibilities and rights as a human? What year will this likely take place?

A18: The consideration of legal rights and responsibilities for robots and AI challenges traditional notions of personhood, rights, responsibilities, and sentience. For robots and AI to be treated similarly to humans under the law, several key characteristics and societal shifts would need to occur:

Characteristics and Considerations:

1. **Consciousness/Sentience:** A robot or AI would need to possess a level of consciousness or sentience comparable to that of a human. It

would need the ability to have subjective experiences, emotions, and perhaps even self-awareness. Without this, granting them human-equivalent rights would be hard to justify.
2. **Autonomy:** The AI or robot would need to operate independently of human intervention, making decisions based on its internal processes rather than strictly following its programming.
3. **Learning and Growth:** Like humans, it might need to have the capability to learn, adapt, and potentially grow in terms of its capabilities, understanding, and experiences.
4. **Accountability:** With rights come responsibilities. If a robot or AI can be granted rights, it should also be accountable for its actions, which poses a challenge. If an AI makes a decision leading to harm, is it the AI's "fault", or does the blame lie with its programmers or operators?
5. **Reproduction and Mortality:** Human rights are deeply tied with notions of life and death. Would an AI have the right to "reproduce" (i.e., create a copy of its software)? Would turning off an AI be considered a form of "killing"?

Timeline and Predictions:

As of my last training cut-off in September 2021, we are nowhere near creating robots or AI with the level of consciousness or sentience that would demand human-equivalent rights. Most AI experts believe that if true artificial general intelligence (AGI) or superintelligent AI ever emerges, it's still decades away, if not more.

However, certain rights and responsibilities might be granted to AI and robots before they achieve anything close to human consciousness:

1. **Legal Personhood:** There are discussions about granting certain robots or AI systems a limited form of "legal personhood." This wouldn't be the same as human rights but would be more akin to the rights and responsibilities of corporations.
2. **Liability:** As AI systems become more integrated into society, there will be an increasing need to determine liability for their actions, especially when human safety is at stake. This could lead to legal frameworks where AI systems have certain "responsibilities."
3. **Protection Rights:** As robots become more advanced, there might be considerations about protecting them from damage or misuse, not because they're sentient, but because of their economic and societal value.

Given the current pace of development, it might be decades before we see substantial legal considerations about AI rights, and it could be well into the latter half of the 21st century before there's any serious contemplation of granting AI human-equivalent rights — if ever. Such developments would also hinge on significant philosophical, ethical, and societal shifts in our understanding of consciousness, rights, and what it means to be a "person."

11.3 Discussion and Conclusion

According to ChatGPT's answers, the world of tourism in 2068 will be driven by a few key words: personalisation (see A1–A8, A13, A16), experiences (A1, A3–A11, A13–A18), sustainability (A1–A9, A11–A16) and well-being (A4–A7, A9, A13–A16). Regardless of whether ChatGPT talks about transportation, accommodation, F&B, visitor attractions, travel motives, corporate or destination competitiveness, these themes are mentioned in one form or another. The answers of ChatGPT are quite relevant because previous studies have already elaborated on the contribution of robots, AI and automation technologies in creating memorable and personalised experiences for customers/tourists (Ivanov, 2019a; Naumov, 2019) that contribute to their well-being (Gani et al., 2023). Additionally, studies have confirmed the roles of robots, AI and automation technologies for the sustainable development of companies/organisations and destinations (Ivanov et al., 2023; Sætra, 2023). Therefore, ChatGPT's responses are in line with previous studies. At the same time, ChatGPT emphasises the authenticity of the tourist experience (A3, A5–A7, A13, A14) which is supported by previous studies (Guerra et al., 2022; Moore et al., 2021). However, the use of automation technologies may hurt the authenticity of services, as emphasised by Seyitoğlu (2021). Therefore, it is not surprising that ChatGPT mentions in its answer (A6) digital detox and disconnection as potential travel motives (Jiang & Balaji, 2022; Stäheli & Stoltenberg, 2022) which might be very important in a robotised society for the mental well-being of people.

It should be noted that the answers provided by ChatGPT, although very relevant and correct, are also stereotyped and follow similar patterns in terms of ideas and recommendations. Often, the answers are nearly identical – e.g. recommending health and wellness tourism for various destinations. Nevertheless, ChatGPT showed creativity in its answers to some questions. When asked about the five leading destinations for international tourists in 2068, ChatGPT mentioned space destinations (Moon, orbital hotels and Mars), VR and AR, underwater cities and resorts, less explored regions such as Antarctica, Greenland or remote islands, and eco and sustainable destinations, instead of mentioning only five specific countries as the authors implied in their prompt. Space tourism (Toivonen, 2021; Webber, 2013) and trips to underwater cities and resorts will be the two types of tourism that robots, AI and automation will make feasible. Although many studies focus on the metaverse and the virtual experiences of people (Buhalis et al., 2023; Israel et al., 2019; Yung & Khoo-Lattimore, 2019), one can argue whether these virtual 'tourists' are tourists at all if they do not leave their usual environment and travel to visit a destination which is the core requirement for a traveller to be considered as a visitor (see UNWTO, n.d.). However, the metaverse, VR and AR will play a vital role at every stage of the tourist trip – from the inspiration and the booking through the travel and the stay to the post-stay behaviour. Therefore, yet again, ChatGPT provided a relevant answer.

Looking at robots as tourists (A17), ChatGPT outlined that to be considered tourists, robots need advanced consciousness to be able to 'appreciate, experience, and seek out new experiences'. Therefore, ChatGPT reported that consciousness is a precondition of the tourism experience. This is a novel statement because in a previous study on robots as consumers, Ivanov and Webster (2017a) raised no such precondition. The authors proved that from an accounting point of view, there is absolutely no difference if a purchase is made by a human or by a robot; hence, robots can be considered as consumers. However, ChatGPT's answer went beyond the accounting perspective and considered the psychological aspects of the consumption process in tourism after the actual purchase similar to other studies (Câmara et al., 2023; Prebensen et al., 2014). Additionally, ChatGPT paid attention to the specific facilities and services that robotic tourists may need, similar to the study of Ivanov and Webster (2017b). Moreover, ChatGPT raised the reasonable question about the economic model to sustain robotic tourists. It asks the reasonable question 'Who pays for the robot's experience?'. In his study on tourism beyond humans, Ivanov (2019b) outlines that a specific characteristic of non-human travellers (robots, pets, toys) is that their trip is booked and paid for by their human owner. Hence, the tourist experience of robots is to be paid for by humans. However, if robots are granted some rights (Gunkel, 2018; Schwitzgebel, 2023; Tigard, 2023), they might have the financial ability to pay for their own trips.

One thing that seems to be missing from ChatGPT's responses is a consideration of trends to counteract the current reality but often in slow and subtle ways, specifically demographic trends. In response to the question about future source markets, A10 makes references to populations that currently have high population levels (China, Russia and Europe), but fails to note the impact of declining demographics into the future. The answer of ChatGPT suggests that population levels will remain high relative to other source markets, but it seems to ignore the rather sizeable impact that the population declines over the next few decades will have on the number and purchasing power of tourists from these markets in the coming decades. While there is ample evidence that there is a long-term problem in much of the world with declining populations (Webster, 2021), there is serious concern specifically in terms of China (Gu et al., 2021; Jiang et al., 2013), Russia (Shcherbakova, 2022) and Europe (Vignoli et al., 2020). These populations may be large relative to other regions/countries, but with quantitative declines and qualitative changes these source markets may require a different type of tourism product. For example, in Russia specifically, the death rates of middle-aged and older males are quite high (Shcherbakova, 2022), so that the future population of Russia will be unusually female and old. As for China, the population has not been at replacement levels since the 1990s (Jiang *et al.*, 2013), so that in a few more decades, its population will also be significantly lower and disproportionately male due to selective abortions (Wang & Jiang, 2022). These demographic trends are very important because as already discussed in Chapter 1, the declining population is one of the major drivers of automation (Webster, 2021), including in the tourism context (Webster & Ivanov, 2020).

Another missing issue is that in A12 ChatGPT makes light of travel behaviour change. Since the climate and weather patterns may change so drastically, it is also likely that experiences with air travel iturbulence may entice many travellers to change their preferred mode of transportation or change their emotional relationship with travel. While ChatGPT notes the practical elements of changes in behaviour, it seems to overlook the impact that such changes may have on humans' experiences of anxiety with travel. Such an increase in anxiety may lead to different travel behaviours and different attitudes towards travel, in general.

In conclusion, ChatGPT depicts a very plausible scenario of future tourism, although probably not sufficiently bold considering that 45-year forecasting horizon. Its answers are based on data until 2021 and its forecasts seem largely as linear extrapolations of what is already known (e.g. accessibility, personalisation, experiences). Nevertheless, the interview provides some very useful insights that confirm the applicability of AI in forecasting tourism's futures.

In addition, it seems that ChatGPT, much like any entity making forecasts about the future, is part of a zeitgeist, basing its predictions and using the language based on the data and language from its own time period. In A18, ChatGPT notes that 'my last training cut-off in September 2021', illustrating the timing of the data being input into its training dataset. It would seem that the technology is using the language and philosophy that are prevalent in our age and making projections based on them, so its predictions seem to be an artefact of our times.

References

Buhalis, D., Leung, D. and Lin, M. (2023) Metaverse as a disruptive technology revolutionising tourism management and marketing. *Tourism Management* 97, 104724. https://doi.org/10.1016/j.tourman.2023.104724

Câmara, E., Pocinho, M., Agapito, D. and Jesus, S.N. de (2023) Meaningful experiences in tourism: A systematic review of psychological constructs. *European Journal of Tourism Research* 34, 3403. https://doi.org/10.54055/ejtr.v34i.2964

Carvalho, I. and Ivanov, S. (2024) ChatGPT for tourism: Applications, benefits, and risks. *Tourism Review* 79 (2), 290–303. https://doi.org/10.1108/TR-02-2023-0088

Dogru, T., Line, N., Mody, M., Hanks, L., Abbott, J., Acikgoz, F., Assaf, A., Bakir, S., Berbekova, A., Bilgihan, A., Dalton, A., Erkmen, E., Geronasso, M., Gomez, D., Graves, S., Iskender, A., Ivanov, S., Kizildag, M., Lee, M. ... Zhang, T. (2023) Generative artificial intelligence in the hospitality and tourism industry: Developing a framework for future research. *Journal of Hospitality & Tourism Research*. https://doi.org/10.1177/10963480231188663

Dwivedi, Y.K., Kshetri, N., Hughes, L., Slade, E.L., Jeyaraj, A., Kar, A.K. ... Wright, R. (2023) 'So what if ChatGPT wrote it?' Multidisciplinary perspectives on opportunities, challenges and implications of generative conversational AI for research, practice and policy. *International Journal of Information Management* 71, 102642. https://doi.org/10.1016/j.ijinfomgt.2023.102642

Fusté-Forné, F. and Orea-Giner, A. (2023) Gastronomy in tourism management and marketing: An interview with ChatGPT. *ROBONOMICS: The Journal of the Automated Economy* 4, 42. https://journal.robonomics.science/index.php/rj/article/view/42

Gani, M.O., Roy, H., Faroque, A.R., Rahman, M.S. and Munawara, M. (2023) Smart tourism technologies for the psychological well-being of tourists: A Bangladesh perspective. *Journal of Hospitality and Tourism Insights* In press. https://doi.org/10.1108/JHTI-06-2022-0239

Gu, D., Andreev, K. and Dupre, M.E. (2021) Major trends in population growth around the world. *China CDC Weekly* 3 (28), 604–613. doi: 10.46234/ccdcw2021.160

Guerra, T., Moreno, P., Araújo de Almeida, A.S. and Vitorino, L. (2022) Authenticity in industrial heritage tourism sites: Local community perspectives. *European Journal of Tourism Research* 32, 3208. https://doi.org/10.54055/ejtr.v32i.2379

Gunkel, D.J. (2018) *Robot Rights*. The MIT Press.

Israel, K., Tscheulin, D. and Zerres, C. (2019) Virtual reality in the hotel industry: Assessing the acceptance of immersive hotel presentation. *European Journal of Tourism Research* 21, 5–22. https://doi.org/10.54055/ejtr.v21i.355

Ivanov, S. (2019a) Ultimate transformation: How will automation technologies disrupt the travel, tourism and hospitality industries? *Zeitschrift für Tourismuswissenschaft* 11 (1), 25–43.

Ivanov, S. (2019b) Tourism beyond humans: Robots, pets and teddy bears. In G. Rafailova and S. Marinov (eds) *Tourism and Intercultural Communication and Innovations* (pp. 12–30). Cambridge Scholars Publishing.

Ivanov, S. and Webster, C. (2017a) The robot as a consumer: A research agenda. In *Proceedings of the 'Marketing: Experience and Perspectives' Conference*, 29–30 June, University of Economics-Varna, Bulgaria, pp. 71–79. SSRN. URL: http://ssrn.com/abstract=2960824

Ivanov, S. and Webster, C. (2017b) Designing robot-friendly hospitality facilities. In *Proceedings of the Scientific Conference 'Tourism. Innovations. Strategies'*, 13–14 October, Bourgas, Bulgaria, pp. 74–81. SSRN. URL: http://ssrn.com/abstract=3053206

Ivanov, S., Duglio, S. and Beltramo, R. (2023) Robots in tourism and sustainable development goals: Tourism Agenda 2030 Perspective article. *Tourism Review* 78 (2), 352–360. https://doi.org/10.1108/TR-08-2022-0404

Jiang, Q., Li, S. and Feldman M.W. (2013) China's population policy at the crossroads: Social impacts and prospects. *Asian Journal of Social Science* 41 (2), 193–218. https://doi.org/10.1163/15685314-12341298.

Jiang, Y. and Balaji, M.S. (2022) Getting unwired: What drives travellers to take a digital detox holiday? *Tourism Recreation Research* 47 (5–6), 453–469. https://doi.org/10.1080/02508281.2021.1889801

Moore, K., Buchmann, A., Månsson, M. and Fisher, D. (2021) Authenticity in tourism theory and experience. Practically indispensable and theoretically mischievous? *Annals of Tourism Research* 89, 103208. https://doi.org/10.1016/j.annals.2021.103208

Naumov, N. (2019) The impact of robots, artificial intelligence, and service automation on service quality and service experience in hospitality. In S. Ivanov and C. Webster (eds) *Robots, Artificial Intelligence, and Service Automation in Travel, Tourism and Hospitality* (pp. 123–133). Emerald Publishing.

Prebensen, N.K., Chen, J.S. and Uysal, M.S. (eds) (2014) *Creating Experience Value in Tourism*. CAB International.

Sætra, H.S. (2023) *Technology and Sustainable Development: The Promise and Pitfalls of Techno-Solutionism*. Taylor & Francis.

Schwitzgebel, E. (2023) The full rights dilemma for AI systems of debatable moral personhood. *ROBONOMICS: The Journal of the Automated Economy* 4, 32. https://journal.robonomics.science/index.php/rj/article/view/32

Shcherbakova, E.M. (2022) Population dynamics in Russia in the context of global trends. *Studies on Russian Economic Development* 33 (4), 409–421. https://doi.org/10.1134/S1075700722040098

Seyitoğlu, F. (2021) Automation vs authenticity in services. *ROBONOMICS: The Journal of the Automated Economy* 2, 20. https://journal.robonomics.science/index.php/rj/article/view/20

Soliman, M. and Al Balushi, M.K. (2023) Unveiling destination evangelism through generative AI tools. *ROBONOMICS: The Journal of the Automated Economy* 4, 54. https://journal.robonomics.science/index.php/rj/article/view/54

Soliman, M., Al-Shanfari, L.S. and Gulvady, S. (2023) Sensory marketing and accessible tourism: An AI-generated article. *ROBONOMICS: The Journal of the Automated Economy* 4, 53. https://journal.robonomics.science/index.php/rj/article/view/53

Stäheli, U. and Stoltenberg, L. (2022) Digital detox tourism: Practices of analogization. *New Media & Society*. https://doi.org/10.1177/14614448211072808.

Tarbit, J., Wirtz, J., Kunz, W. and Hartley, N. (2023) Interpretation of corporate digital responsibility risks and concerns by automated service technologies: An AI co-created article. *ROBONOMICS: The Journal of the Automated Economy* 4, 52. https://journal.robonomics.science/index.php/rj/article/view/52

Tigard, D. (2023) On respect for robots. *ROBONOMICS: The Journal of the Automated Economy* 4, 37. https://journal.robonomics.science/index.php/rj/article/view/37

Toivonen, A. (2021) *Sustainable Space Tourism: An Introduction*. Channel View Publications.

UNWTO (n.d.) *Glossary of Tourism Terms*. UNWTO. https://www.unwto.org/glossary-tourism-terms

Vignoli, D., Guetto, R., Bazzani, G., Pirani, E. and Minello, A. (2020) A reflection on economic uncertainty and fertility in Europe: The narrative framework. *Genus* 76, 28. https://doi.org/10.1186/s41118-020-00094-3

Wang, T. and Jiang, Q. (2022) Recent trend and correlates of induced abortion in China: Evidence from the 2017 China Fertility Survey. *BMC Women's Health* 22, 469. https://doi.org/10.1186/s12905-022-02074-5

Webber, D. (2013) Space tourism: Its history, future and importance. *Acta Astronautica* 92 (2), 138–143. https://doi.org/10.1016/j.actaastro.2012.04.038

Webster, C. (2021) Demography as a driver of robonomics. *ROBONOMICS: The Journal of the Automated Economy* 1, 12. https://journal.robonomics.science/index.php/rj/article/view/12

Webster, C. and Ivanov, S. (2020) Demographic change as a driver for tourism automation. *Journal of Tourism Futures* 6 (3), 263–270. https://doi.org/10.1108/JTF-10-2019-0109

Yung, R. and Khoo-Lattimore, C. (2019) New realities: A systematic literature review on virtual reality and augmented reality in tourism research. *Current Issues in Tourism* 22 (17), 2056–2081. https://doi.org/10.1080/13683500.2017.1417359

Conclusion

The History of the Future: Robonomics and Tourism

Craig Webster and Stanislav Ivanov

> The real problem of humanity is the following: We have paleolithic emotions, medieval institutions and godlike technology. Edward O. Wilson quoted in Tristan Harris (2019) Our brains are no match for our technology. *New York Times* https://www.nytimes.com/2019/12/05/opinion/digital-technology-brain.html

You have been reading a history of the future, a future in which automation serves humans and provides for unrivalled amounts of material prosperity that should allow humans to live in comfort as never before. But there will have to be social transformations that are consistent with and supportive of the new economic order in which we will live. Robonomics should provide the material basis of Marx and Engels's predicted communist utopia, at least in terms of the ability of the system to create wealth to support humans. It is also the culmination of the capitalist's dream of efficiency and productivity, although lacking a clear mechanism to create market demand.

While robonomics increasingly undermines the notion of the labour theory of value by removing human labour from the production of wealth, it brings up something that could become a major institutional crisis. There must be a way to make a system run by machines that serves the needs of humanity. This is where the transformation of the social institutions will need to be made in order for the basis of the productive capacity to be a sustainable force that will serve humanity long into the future. Humans and technological innovations will have to co-create and re-create social institutions that are both consistent with the primate brains of humans and the godlike technological capabilities that are the fruits of millennia of human creativity, ingenuity and hard work. A great challenge will be to quickly create effective and sustainable social institutions to ensure that humans can benefit from the productive and efficient capabilities that technologies afford us. With the massive capabilities of technologies, humans can live in a world of wealth and abundance like never before, but only if social institutions can be engineered that enable

Paleolithic and space age technologies. As they are rumoured to say at NASA, 'failure is not an option'.

We hope that humanity continues to march into the future, sometimes clearly in a successful way, even with occasional failures. No other species known has been so successful and has created so much. Backsliding into tribalism, poverty and savage violence is undesirable, although occasional backsliding is possible. We hope that humanity can meet the challenges of robonomics as successfully as it has met every challenge before.

This book has discussed robonomics and the role of tourism in a robonomic world. The chapters explain more fully how the political economy of tourism will function once we are more deeply into this Fourth Industrial Revolution. While some things in tourism may change, others will stay the same. There will still be hotels, railways, airlines and the other assorted infrastructure we currently associate with tourism. However, there will be a transformation in terms of how things are done because humans will largely be removed from the labour pool that currently supplies the goods and services that are used in tourism-related industries. In front-of-house hospitality, robots will cook and deliver food to customers and artificial intelligence (AI) will use predictive technologies to cater to specific customers' needs. Back of house in hotels and restaurants, payroll, payments, legal agreements, insurance and purchasing from suppliers will be automated and optimised by AI technologies. The end result should be tourism that is efficient, inexpensive and yet still customised to the wants and needs of specific customers because of the accessibility of data.

While some aspects of tourism may stay the same, tourism will play a different role than it had in previous generations. The financial and productive foundations upon which tourism and hospitality take place will have been transformed, since few humans will work, most humans will be dependent on universal basic income and much of the labour in the tourism industry will be automated. In some ways, tourism will be democratised but it will also be used as a social and political tool to ensure that there is a sense of social and political stability, so that the robonomic system can go forth in a sustainable way into the future.

We are on the cusp of a robonomic world, since we have moved quickly from science fiction to science fact. We hope that this book and the contributions of the authors of the book have given you some insight into how the future is unfolding and giving you the ability to plan and prepare for the new social, economic and political reality we face. While there will be political, economic and social turbulence in the new robonomic future, the likely outcome will lead to a better future for humanity. This book, as the culmination of efforts of scholars from around the world, should help you, the reader, do as the Boy Scout motto proclaims, 'Be Prepared'.

Index

Accommodation: 14, 34, 35, 116, 119–120, 127, 153, 158, 163, 170, 217, 223, 224, 226–228, 231, 234, 238, 240, 244, 248, 251
Artificial autonomous agent: 45, 46, 49, 50, 66, 96, 97, 103–105, 107, 138, 140–144, 210, 212, 217
Artificial intelligence: 28, 44, 69, 85, 103, 114, 115, 122, 133, 149, 170, 181, 185, 191, 211, 223, 233, 257
Attitudes: 46, 55, 74, 108, 140, 198, 253
Automated decision-making: 46, 138
    Human-in-the-loop: 50
    Human-on-the-loop: 50, 138
    Human-out-of-the-loop: 50, 51, 133, 138
    AA-in-the-loop: 50, 51
    AA-on-the-loop: 50, 51
Automation: 26, 28, 30, 36–39, 43–46, 48–49, 51–60, 65–78, 80, 81, 85–87, 91–94, 103–105, 108, 110, 114, 115, 117, 119–127, 133–140, 142–144, 148–150, 152–163, 166, 170, 178, 192, 210–213, 215–220, 223–225, 229, 232–235, 251, 253, 256
    Ironies of automation: 67–68, 137
    Partial automation: 137, 138, 144
    Full automation: 134, 135, 137, 138
    Task automatability: 134, 135

Battery: 70, 185, 211, 219
Behavioural control: 53, 80, 105
Biosecurity: 46, 56
Birth control: 47, 94, 95, 107

ChatGPT: 3, 13, 223, 224, 251–253
Competitive advantage: 46, 49, 52, 56, 68, 105, 107, 140, 142, 210, 214, 215, 218, 232

Demography: 46, 54, 79, 92, 97, 107, 114
    Demographic crisis: 56, 59, 75, 94, 95
    Overpopulation: 43, 73, 75, 105

Education: 32, 47, 78, 85, 88–90, 106, 108, 149–151, 166, 202, 217, 230, 231, 235, 236
Efficiency: 26, 28, 46, 51–53, 58, 59, 66–68, 80, 86, 90, 108, 115, 117, 123–125, 137, 150, 156–158, 166, 174, 183, 210–212, 216, 217, 224–227, 229, 232, 234, 256
Employment: 28, 45–47, 52, 55, 56, 74–76, 79, 86–88, 90, 91, 95, 104–107, 109, 160, 210, 213, 217
    Technological unemployment: 72–77, 86, 88, 92, 95, 106, 110
    Unemployment: 45, 47, 60, 73–75, 77, 87, 105, 210, 211
    Labour market: 44, 54, 56, 75, 91
Energy: 5, 17, 21, 43, 69, 109, 118, 164, 179, 184, 190, 210–214, 216–219, 227, 230, 234, 248
Entertainment: 35, 47, 89–90, 106, 109, 110, 120, 150, 151, 161, 163, 166, 237
Experiences: 26, 35, 89, 90, 103, 114, 116, 119, 120, 122, 124, 125, 127, 134–136, 138–141, 144, 148–154,

156, 159, 160, 163–166, 172, 174, 186, 192, 196, 198, 201, 218, 224–238, 241–253
Experience design: 133, 138, 140, 150, 159, 160, 230
CREATIVE framework: 150–152, 154

Financial management: 142–143
Functional illiteracy: 47, 77–78, 106
Future: 18, 21, 26–31, 33, 37–39, 45, 48, 49, 53, 54, 56, 57, 59, 68–70, 72, 85, 88, 92, 95, 96, 103, 111, 114, 119, 121, 128, 133–136, 138–141, 143, 144, 148, 150, 152–154, 157, 159, 161, 164, 177, 181, 183, 191–195, 197–201, 203–205, 209–210, 212, 219, 220, 223–226, 228–237, 239, 240, 242, 244, 248, 252, 253, 256, 257
Trajectories: 114–117, 119, 122–124, 127

Global citizenship: 47, 72, 105
Global government: 47, 72, 92, 105

Health: 30, 31, 43, 47, 48, 53, 70, 71, 73, 80, 81, 90, 92, 95, 117, 143, 144, 153, 155, 157, 164, 191, 192, 199–203, 216, 228–234, 236, 240, 241, 243, 251
Hotel: 7, 9–12, 14, 16–23, 31, 32, 34–36, 38, 50, 57, 119, 135, 136, 141, 142, 148, 149, 159–161, 163, 164, 171, 172, 175–177, 179–181, 184, 224
Housekeeping: 116, 120, 134, 138, 159, 171, 177, 234
Human augmentation: 71, 79, 142
Human rights: 72, 77, 250
Redefinition of human rights: 47, 90, 94–95, 107
Human values: 79, 106, 193, 203

Industrial Revolution: 43–45, 190–192
Fourth Industrial Revolution: 44, 104, 111, 190, 191, 203, 204, 257
Innovation: 28, 29, 109, 114, 116, 121, 122, 126, 127, 148, 166, 177, 184, 191, 213, 225, 226, 228, 232, 233, 235, 238, 240, 244, 246, 256
Output-based innovation: 150, 151, 153, 156–160, 162–165
Experience-based innovation: 152, 153, 157–158, 160, 163, 165, 166
Systemic innovation: 150, 154, 158, 160–161, 164–166
Process-based innovation: 150–156, 159, 161–162, 164

Keynes, J.: 74
Kitchen: 38, 135, 136, 161, 175, 176

Law, R.: 114, 153, 154, 156, 159, 160, 162, 164

Marketing management: 141–142
AI2AI marketing: 105, 107, 141

Operations management: 66, 134, 138–140

Polanyi's paradox: 67
Political instability: 47, 72, 76–77, 106
Politics: 54–55, 71, 89, 95
Privacy: 53, 55, 80–81, 105, 119, 127, 161, 167, 185, 210, 218, 236
Productivity: 43, 55, 58, 59, 66, 68, 74, 104, 107, 125, 138, 211–213, 216, 256
Profitability: 58, 86, 105–107, 151, 211, 213, 215, 216

Quality of life: 43, 47, 70–72, 87, 89, 91, 95, 105, 108, 110, 112, 211, 213, 218, 220

Regulation: 72, 227
Restaurant: 6, 8, 10, 11, 17, 18, 20, 22, 23, 33–36, 38, 67, 68, 136, 137, 156, 164, 172, 180, 183, 217, 218
Robonomics
Definition: 45
Drivers of: 53–56
Principles of: 48–53
Benefits of: 43, 45, 47, 48, 52, 59, 60
Challenges of: 45, 48, 52, 60
Solutions to the challenges of: 45, 47, 85–97

Robot: 16, 20, 36, 38, 47, 51, 53, 57, 59, 67, 68, 73, 91–93, 96–97, 104, 105, 107, 120, 124, 127, 134, 135, 137, 139, 140, 142, 148, 149, 162, 171–186, 191–193, 195–205, 223, 248–250, 252
  Definition: 45
  Industrial robots: 48, 68
  Social robots: 48, 51, 135
  Service robots: 53, 58, 65, 103, 123, 140, 180–185, 233
Robot rights: 96, 249–251
Robot-friendly hospitality facilities: 104, 140, 170–186
  Design of: 174–184
Roman Empire: 110

Sex: 53, 80, 90, 171, 190, 192, 196–199, 202
Sustainability: 28, 29, 47, 69–71, 86, 105, 140, 141, 185, 209–211, 214, 216, 219, 220, 225, 227–235, 238, 240, 242, 243, 247, 251
Sustainable Development Goals (SDGs): 140, 209, 216, 220

Taxation: 47, 68
  Robot-based taxation: 47, 92–94, 107, 124, 127, 223
  Tax policy: 47, 88, 106

Tourism: 26–28, 32–39, 45, 47, 54, 73, 89–90, 103–112, 114–117, 120, 124–128, 133–145, 148, 151, 152, 157, 163, 170, 174, 179, 190, 192, 195–198, 200, 201, 209, 210, 216–220, 223–253
Tourist: 5, 21, 26, 29, 32–36, 103–106, 141, 144, 149, 150, 152–155, 160, 199, 202, 203, 224, 234, 237–240, 243, 245, 248, 251, 252
Transport: 26–28, 31, 32, 48, 116–119, 127, 190, 213, 223, 225, 226, 228
  Autonomous vehicles: 28, 37, 45, 48, 50, 69, 77, 81, 118, 119, 148, 149, 172, 182, 185, 225, 234

Überveillance: 53, 55, 80–81, 105, 210, 215–218, 220
Universal basic income (UBI): 14, 47, 72, 75, 77, 78, 90–92, 94, 95, 97, 107, 109, 111, 213, 214, 217, 220, 257

Visitor attractions: 120–122, 223, 229, 230, 234, 251
Volunteering: 47, 90, 95, 106, 109, 213

Wellness: 190–205, 229, 230–233, 236, 241, 243, 245, 247, 251

www.ingramcontent.com/pod-product-compliance
Ingram Content Group UK Ltd.
Pitfield, Milton Keynes, MK11 3LW, UK
UKHW021836140426
5217IPUK00021B/1482

For Product Safety Concerns and Information please contact our EU Authorised Representative:

Easy Access System Europe

Mustamäe tee 50

10621 Tallinn

Estonia

gpsr.requests@easproject.com